Palgrave Studies in the History of Science and Technology

Series Editors

James Rodger Fleming
Colby College
Waterville, ME, USA

Roger D. Launius
Auburn, AL, USA

Designed to bridge the gap between the history of science and the history of technology, this series publishes the best new work by promising and accomplished authors in both areas. In particular, it offers historical perspectives on issues of current and ongoing concern, provides international and global perspectives on scientific issues, and encourages productive communication between historians and practicing scientists.

Mikael Hård

Microhistories of Technology

Making the World

palgrave
macmillan

Mikael Hård
Institute of History
Technical University of Darmstadt
Darmstadt, Germany

ISSN 2730-972X ISSN 2730-9738 (electronic)
Palgrave Studies in the History of Science and Technology
ISBN 978-3-031-22812-4 ISBN 978-3-031-22813-1 (eBook)
https://doi.org/10.1007/978-3-031-22813-1

Cover illustration: bauhaus1000 / Getty Images

This Palgrave Macmillan imprint is published by the registered company Springer Nature Switzerland AG.
The registered company address is: Gewerbestrasse 11, 6330 Cham, Switzerland

Preface

As a young doctoral student in the 1980s, I was convinced that the history of the nineteenth and twentieth centuries could best be understood via concepts like *industrialization*, *urbanization*, and *rationalization*. The research sources I used for my PhD dissertation included engineering formulae and technical drawings. And there was no doubt in my mind that Sigfried Giedion had summarized modern history in the three-word title of his most famous book: *Mechanization Takes Command*. Only later did it dawn on me that history is much more complex, and frequently contradictory. In time, I understood that the past cannot be reduced to one-dimensional processes such as mechanization.

When, two decades later, my focus shifted from Europe to other parts of the word, I began working with another one-dimensional concept: *globalization*. In standard historical narratives, globalization has been portrayed as an unstoppable force that flattens all obstacles in its path; modern technology is usually seen as globalization's constant companion. For example, in the second half of the nineteenth century, steamships, telegraph lines, and Gatling guns are said to have paved the way for colonialism, one of the defining phenomena of that time in history. Another example: During the second half of the twentieth century, shipping containers and computer networks purportedly pushed the world further in the direction of globalization.

Increasingly uneasy with such overarching, *macro*historical interpretations, I asked myself if a *micro*historical approach might help us see the world from a new angle. What if we were to tell stories from "below" rather than from "above"—focusing on ordinary people and their experiences rather than anonymous structures? After all, peoples in Asia, Africa, and Latin America have created their own forms of material culture and used imported technologies in different ways. An even more exciting prospect: Could research on the Global South help to transcend received notions of history? Could this investigation contribute to new paradigms in the field of history of technology—beyond simplistic notions like technology "diffusion" and "transfer"? These are the questions which guided me when conceptualizing this book.

To view Western technology as an invincible juggernaut is to leave no room for local initiatives and culturally specific technologies. It is time to question assumptions about ever-increasing global uniformity. After all, the world displays wildly divergent technologies. Rather than seeking universal solutions—let alone pushing for them—we need to acknowledge the varied technologies made by human beings across the world. As a historian leading an international team of researchers in a project called Global-HoT (A Global History of Technology, 1850–2000), my contribution to this endeavor is to uncover and recount representative microhistories—rather than to perform macroanalyses. In this book, we have gathered stories of how people in various corners of the world challenged the production techniques and products brought by globalization. Exercising their autonomy and freedom, creative people from all walks of life selectively modified, adopted, or rejected modern machines, tools, and gadgets. The spread of so-called modern technologies did not erase artisanal production methods and traditional tools. Often, the new and the old coexisted. Cultural encounters between the Global North and the Global South have given rise to countless innovative solutions, many of which have gone unnarrated.

By emphasizing the richness of material cultures beyond Europe and North America, the team and I illustrate that Western societies do not have a monopoly on novelty. And by bringing to light the technologies of skillful people around the globe, we show how inventive individuals and groups have shaped their lives and their communities. Each chapter in this volume demonstrates how tenacious people have used traditional, "homemade," and hybrid technologies, maintaining autonomy in a world of corporate systems and global networks.

The global history of technology cannot be reduced to tracing the transfer of technologies "from the West to the rest." On the contrary, we need to embrace the complex reality that local tools and methods have survived despite the onslaught of globalizing forces. In many cases, to meet their needs, indigenous populations creatively combined local tools and methods with modern ones. By redefining "technological development" to include the daily practice of people in various cultural settings, I hope to more fully inform readers of the history of human ingenuity.

The title of this book, *Microhistories of Technology*, is deliberate in its plurality *and* in its reference to being a collection of singular stories. These microhistories were crafted with the aim of unraveling the peculiarities of each regional technological landscape; at the center of each story are the specific knowledge and skills inherent in each local material culture. For example, inhabitants in Northern India employ various technologies, which may be called indigenous or imported, Western or Eastern, modern or traditional. Within this range of possibilities, the combination of solutions each group uses is place- and time-specific; in every case, *the user's* world is singular.

I chose the subtitle of this book to suggest its connection with *Consumers, Tinkerers, Rebels: The People Who Shaped Europe*, a book I coauthored with Ruth Oldenziel and published in 2013—as part of the Palgrave Macmillan book series *Making Europe: Technology and Transformations, 1850–2000*. In contrast to that book and my earlier works, this volume focuses on the interactions of individuals and groups with technology in the Global South, as well as in Central and East Asia. I highlight the practices, struggles, and aspirations of peoples in Africa, Asia, and Latin America. I also apprise readers of the many myths and prejudices typically associated with imperial powers, globalizing forces, and development policies. In doing so, I hope to expand on the new body of literature that examines technology's complex and nuanced role in the modern world. My goal—and indeed the goal of the Global-HoT team—is to inspire readers to rethink the relationship between Western and non-Western material culture.

It has been a pleasure to write this book, and it has been a privilege to discuss its findings with colleagues from so many parts of the world. I have enjoyed the opportunity to craft narratives that bring to the fore previously unknown artisans and marginalized individuals. Some of the actors in this book are local technicians and craftspeople who learned to adapt "foreign" technologies to their own—and their communities'—needs. Others are villagers and "slum" dwellers who singlehandedly made products to sell locally. Still others are people who made their own products in an effort to retain their freedom of choice in a world increasingly dominated by commercial consumer goods. Although local and "micro," these people's stories deserve to be told.

Acknowledgments

This book is the outcome of a collective effort. That said, the cover of this book cites my name as the sole author. Indeed, I wrote all the chapters, and I take full responsibility for any and all mistakes in the text. However, the material herewith is based on research that I and other scholars conducted within the framework of the research project "A Global History of Technology, 1850–2000" (Global-HoT). This five-year undertaking began in 2017, and it has been generously funded by the European Research Council as a part of the Research and Innovation Program *Horizon 2020* of the European Union (*ERC Advanced Grant No. 742631*). This research grant has also paid for language editing and open-access publishing. My writing has also benefited from intense interaction with fellows of the PhD program "Urban Infrastructures in Transition: The Case of African Cities," which was financed by the Hans Böckler Foundation from 2014 to 2019 (*Promotionskolleg PK039*). In addition, it would not have been possible to carry out the foundational work for the book if I and my colleagues had not received dedicated support from the Technical University of Darmstadt (TU Darmstadt), a public institution of the German State of Hesse. I acknowledge the central-administration staff members of TU Darmstadt, the university library, and my long-time secretary, Iris Ohlrogge. Simon Bihr helped me and the rest of the ERC team to solve all kinds of IT-related problems.

Indeed, it would be presumptuous for anyone to claim to singlehandedly write a truly global history of technology; close collaboration with researchers from various continents is a requirement. Members of the research clusters mentioned above contributed empirical material, their language skills, and their competence in communicating cultural phenomena. These collaborations have yielded scanned archival material from distant archives and extensive references to secondary literature. The material in this book has also benefitted enormously from critical and creative seminar discussions that took place under the auspices of the European Research Council project and the Böckler program. This exchange and these discussions have taught me to constantly question—and correct—my own perspective, my worldview, and my implicit prejudices.

I have conducted primary research in archives and libraries in the United Kingdom, India, and East Africa; my colleagues have carried out primary research in many European countries, South America, East and West Africa, and various parts of Asia. Twelve of the eighteen members of the research team grew up in countries outside Europe and are fluent in local languages, which include Kiswahili, Uzbek, Korean, Bengali, and Arabic. Most of the researchers have been PhD students or postdocs at TU Darmstadt; others have been associated with Global-HoT on a contractual basis or as assistant researchers.

In terms of the researchers who have contributed to this book, I thank Alejandra Osorio Tarazona, who provided me with literature and sources for the chapter on cooking in South America. Some of the cookbooks quoted in that chapter come from her personal collection. In addition, Alejandra explained the intricacies of various cooking techniques as well as particular Peruvian and Argentine recipes and ingredients. Alejandra also helped me to better understand the specifics of South American material culture.

Youngju Lee has been another invaluable contributor to this book. Without Youngju's assistance, it would have been impossible for me to recount the history of menstruation techniques in South Korea. Youngju was also kind enough to give me access to interviews she had conducted with women in South Korea. Our discussions broadened my knowledge of East Asian history in general—and modern Korean history in particular.

For the chapter on precolonial and colonial West Africa, I benefited enormously from groundwork carried out by David Drengk. David's scanned primary sources, books, and articles enabled me to comprehend a part of the world I have not experienced in person. Our conceptual discussions were crucial to this book; it was David who convinced me to employ the concept of the "technological landscape" in interpreting the West African environment.

Mai Lin Tjoa-Bonatz did a wonderful job of researching information for the first two chapters of this book. A specialist in Indonesian history, Mai Lin unearthed archival material on both gold production and missionary stations. The process of co-authoring two articles with Mai Lin contributed greatly to my cultural understanding of the region and the activities of various groups—Batak headmen, Chinese goldminers, and Christian missionaries among them.

The microhistory of construction work and daily life in Central Asian cities would not have come about without the help of scholars Mariya Petrova and Jonas van der Straeten. Mariya and Jonas carried out the interviews on which much of the chapter is based; these scholars also provided most of the source material. Mariya's language skills and cultural knowledge proved essential during the revision process of the chapter.

The chapter on electricity provision in East Africa was very much inspired by the work of Jonas van der Straeten and Emanuel Lukio Mchome. Both scholars also gave me invaluable—often quite critical—feedback. In addition, Jonas, along with Markus Schertler, provided me with relevant archival material. Markus was one of the ERC program's associate members, although his research was financed by the Deutsche Forschungsgemeinschaft (DFG). Frank Edward helped with access and support during my research in Dar es Salaam, Tanzania.

Animesh Chatterjee and Aloy Buragohain generously provided their comments and suggestions on the chapter on sugar production in India. Jethron Ayumbah Akallah did the same for the chapter on Nairobi's so-called slum areas. At an early stage of the book project, Jethron assisted me in the relevant archives and libraries in Kenya. I am thankful for Animesh, Aloy, and Jethron's willingness to discuss the concepts of "flexible settlement" and "appropriate technology," which I adopted in two of the book's chapters.

Finally, I must mention additional people who have supported and inspired me in the process of writing this book. Jochen Monstadt was instrumental in piquing my interest in technological change in non-European settings. It was Jochen who secured the funding for the aforementioned PhD program on African cities; he also brought in fellowships for African PhDs from the German Academic Exchange Service (DAAD). Without this financial support, I would not have had the opportunity to work with Jethron, Frank, and Emanuel.

In addition, Arne Kaijser, Lisa Friedman, Shorouk El Hariry, Dennis Yazici, Markus Böck, and Wei Wu contributed their comments and ideas; their feedback helped me to refine theories, optimize analyses, and improve my arguments. In particular, Lisa's critical and constructive remarks forced me to rethink my own standpoints and modify my narratives.

Thank you all!

Book Abstract

In most standard historical narratives, globalization is portrayed as an unstoppable force that flattens all obstacles in its path. Modern technology, too, is seen as inexorable: in the nineteenth century, steamships, telegraph lines, and Gatling guns are said to have paved the way for colonialism. Later, shipping containers and computer networks purportedly pulled the planet deeper into a maelstrom of capitalism.

Microhistories of Technology: Making the World tells the alternative story of how people in Africa, Asia, and Latin America challenged globalization, from its products to its production techniques. Creative individuals selectively adopted or rejected modern gadgets, tools, and machines. For example, carpenters in Sumatra during the nineteenth century ignored the wood saws imported by missionaries—choosing to chop down trees with their arch-bladed adzes. Similarly, people in colonial India disregarded competition from capitalist-run sugar plantations, producing their own muscovado for local consumers. And, in Soviet times, inhabitants of Samarkand, Uzbekistan, shunned the new prefabricated, concrete residential buildings, preferring to remain in their existing mud-brick houses.

Through the lens of diversity, this book presents a multifarious global history of technology and material culture. The microhistories in this volume show that the spread of modern technologies did not erase artisanal tools and production methods. Rather, the new and the old often coexisted fruitfully—as when South American home cooks adopted both pre-Hispanic grinding stones and modern electric stoves. The outcome of cultural encounters between old and new has been countless innovative solutions, many of which have gone unnarrated in the history of technology.

By bringing to light the material culture of ordinary people around the world, *Microhistories of Technology* shows how inventive individuals and groups have shaped their own lives. Each chapter demonstrates how the tenacious use of traditional, homemade, and hybrid technologies has helped people to maintain autonomy in a "globalized" world.

Praise for *Microhistories of Technology*

"Ideal for teaching, Hård's eight vivid and illuminating microhistories of "honing local techniques in a global world" will captivate readers, challenging them to think afresh about how globalization works on the ground. Bringing fresh insights into everyday technological choices through the lens of material culture, this fascinating book will tempt readers to further explorations in the history of technology."

—Professor Francesca Bray, *University of Edinburgh*

"Among the efforts to understand globalisation and to write global histories in new and diverse perspectives, this work will stand out for its critical engagement with the question of technology and many presumed directionalities and inevitabilities thereof. It is bound to provoke varied responses and to inspire fresh researches on less explored localities and dimensions—perhaps with even less positing of Europe as the polestar."

—Professor John Bosco Lourdusamy, *Indian Institute of Technology Madras*

"Granitic ideas of Western technology and experts conquering the world come here to a lethal end. Instead, a complex mosaic of micro and local universes of silent, unnamed, and creative contributors to technological landscapes emerges. The book is a feast of ordinary people in the "Global South," making the world through ordinary actions and technologies. The impressive variety of unusual sources highlights historical cases rarely addressed by scholarship."

—Professor Stefania Gallini, *Universidad Nacional de Colombia*

Contents

About the Author

Mikael Hård studied in Uppsala and Gothenburg in Sweden and at Princeton University, NJ, before receiving his PhD in the history of science and ideas from Gothenburg University, Sweden, in 1988.

Between 1988 and 1994, Mikael worked at Gothenburg University and the Berlin Social Science Center (WZB); he also spent 1 year as a research fellow at the Swedish Collegium for Advanced Study (SCAS) in Uppsala. In 1994, Mikael became professor of history of technology at the Norwegian University of Science and Technology (NTNU), and 4 years later he moved on to the Technical University of Darmstadt (TU Darmstadt), Germany, to occupy a professorship with the same denomination.

As a professor at TU Darmstadt, Mikael directed two PhD programs, financed by the DFG, and was an active member of "Tensions of Europe," an international network of historians of technology. In 2017 he received a 5-year grant from the European Research Council (ERC) for a project on which this book is largely based—"A Global History of Technology, 1850–2000" (*ERC AdG 743126*).

Mikael has published widely in high-ranking scientific journals and with academic presses. Among his publications are:

- *Consumers, Tinkerers, Rebels: The People Who Shaped Europe* (Palgrave Macmillan, 2013), written with Ruth Oldenziel
- *Urban Machinery: Inside Modern European Cities* (MIT Press, 2008), co-edited with Thomas J. Misa

- *Hubris and Hybrids: A Cultural History of Technology and Science* (Routledge, 2005), written with Andrew Jamison
- *The Intellectual Appropriation of Technology: Discourses on Modernity, 1900–1939* (MIT Press, 1998), co-edited with Andrew Jamison

CHAPTER 1

Introduction: Honing Local Techniques in a Globalized World

One morning we hiked toward the mystical mountain of wealth and happiness. (According to Dayak legend, a god's son slayed his brother here, and his tears and brain transformed into numerous gold nuggets.) […] Here, the Dayak do not only wash the river sand; they also carry out the important work [of gold] extraction in mines and galleries dug through hard rock containing primary deposits. Particularly astonishing and interesting is the fact that the Indigenous do this without any Chinese or European manager. We are witnessing the gold mines of Gunung Mas, probably the only one of its kind in the Malay archipelago.[1] […]

With great skill, they follow the gold veins impressively far down, break the rock (it should be noted in passing that they employ mallets and chisels of German origin and work in the light of German storm lanterns); transport it in panniers on their backs or along self-made winches to the surface; pound; sift; and wash it; melt it in small, earthen vessels into bars; fashion portions of it into artful jewelry and coveted golden teeth; and sell the rest

[1] Helbig, Karl, "Gold bei den Dajaks," *Ostasiatische Rundschau* 19 (15/16), 1938: 386–390; here: 387. All translations from German into English have been done by the author.

M. Hård, *Microhistories of Technology*, Palgrave Studies in the History of Science and Technology,
https://doi.org/10.1007/978-3-031-22813-1_1

to Malay buyers. Who would expect such a highly developed business among the "savage" Dayak, these most infamous of all headhunters![2]

In 1937, the German geographer and explorer Karl Helbig traversed the island of Borneo on foot. The visit to Gunung Mas was one of the highlights of his strenuous journey. From the village of Tewah, where they had spent the night, Helbig and his party followed an overgrown, neglected road through the jungle to Gunung Mas—"Gold Mountain" in the local Dayak language. The sight of the mountain provoked mixed feelings in Helbig. On the one hand, he lauded the industrious activity he and his small group of companions encountered in the middle of the tropical rainforest: an estimated one thousand Dayak and Chinese people were diligently mining and processing gold deposits. On the other hand, Helbig regretted that earlier gold-mining attempts by European companies had failed. He wrote of his nostalgia for old steam engines and "hefty cogwheels," which were now obscured by "high grass," and he used his camera to document the contrast between the contemporaneous industriousness and the industrial "ruins" of earlier times.[3]

This vignette from the colony of the Dutch East Indies—today's Indonesia—reproduces prejudiced and racist worldviews. Despite Helbig's acknowledging words about Dayak industriousness, the quote from his diary reflects standard European narratives of the time. Two of the most common parables in European historical sources are "uncivilized people" and "underdeveloped countries." Such story lines classified various parts of the world in accordance with a racist, evolutionary logic.[4] According to this biased thinking, European and North American people are typically placed at the top of the pyramid of development; they are followed by Chinese and Indian people, and other members of previous so-called

[2] Helbig, Karl, "Fahrt ins Dunkel. XI: Eine Reise durch Borneo," unpublished travel report (1937/38), in: Roemer- und Pelizaeus-Museum Hildesheim GmbH, Hildesheim, Germany: File No. K. 09: 9.4: "Fotos v. K. Helbig, Java, Borneo, Sumatra (I+II)," p. 19. Dr. Mai Lin Tjoa-Bonatz has been kind enough to provide me with published and archival material from Karl Helbig—including a selection of photographs, in addition to other historical sources about gold processing in colonial Dutch East Indies (today's Indonesia); cf. also Tjoa-Bonatz, Mai Lin, and Mikael Hård, "Creole Objects and Techniques: Gold Mining, Gold Panning and Gold Working in Colonial Indonesia," *Baessler-Archiv* 67, 2021: 67–94.

[3] Helbig, "Gold," 387.

[4] Moon, Suzanne, *Technology and Ethical Idealism: A History of Development in the Netherlands East Indies*. Leiden: CNWS Publications, 2007.

advanced civilizations. At the bottom of the pyramid are what Helbig called "primitive" and other "savage" people.[5]

The quote from Helbig also illustrates the place of tools and machines on this imagined evolutionary and racist ladder.[6] Helbig held to the idea that Western technology represented the zenith of historical development. This idea was particularly ironic: in fact, the rainforest had reclaimed the railroad tracks and bridges built by a Dutch company at the turn of the twentieth century. Also gone was the "heavy machinery" the company had installed, as well as the "big industrial plants" it had erected.[7] The observation that the Dayak miners made use of German implements did not challenge the dichotomy in Helbig's mind between "ancient manual work" and "modern machines."[8]

Although he subscribed to this Eurocentric, colonialist narrative, Helbig appreciated the "simple methods" that the "humble" Dayak profitably employed.[9] As indicated by an extensive photo collection, Helbig appears to have been genuinely impressed by the implements used by the indigenous gold seekers: wooden pans (***dulang-dulang***), bamboo ladders, iron mortars, and clay pots. Further, Helbig was convinced that any European pit foreman would "wonder" how efficiently the workers were able to deploy such artifacts. When it came to the manufacture of gold items, Helbig, similarly, expressed surprise at the beautiful jewelry the Dayak goldsmiths were able to design with their "most primitive tools."[10]

From a contemporary, postcolonial perspective, it is evident that Helbig combined Eurocentric, evolutionary thinking with genuine admiration for the achievements of the Dayak. In his seminal book *Orientalism*, Edward Said provides critical insights into the tendency of European commentators—including philologists, archaeologists, and historians—to romanticize faraway cultures.[11] Unfortunately, the West's fascination with the "Orient" does not mean that Western commentators regard Eastern cultures as equal—or as having reached the same level of development. On

[5] Helbig, "Gold," 388.

[6] Cf. Adas, Michael, *Machines as the Measure of Men: Science, Technology, and Ideologies of Western Dominance*. Ithaca, NY: Cornell University Press, 1989.

[7] Helbig, "Gold," 387.

[8] Helbig, Karl, *Eine Durchquerung der Insel Borneo (Kalimantan): Nach den Tagebüchern aus dem Jahre 1937*, Vol. 2. Berlin: Dietrich Reimer, 1982, pp. 364, 366.

[9] Ibid., p. 366.

[10] Ibid., pp. 366, 437.

[11] Said, Edward, *Orientalism*. New York: Pantheon, 1978.

the contrary, as Said notes, the "enthusiasm" for "the Other" cannot be separated from questions "of power, of domination."[12] In line with such postmodern thinking, I try, in this book, to interpret past ideologies and practices within a critical, anti-imperialist framework.[13]

Coexistence and Culture

Helbig's writings and photos serve as a point of departure for the microhistories I present in this book. Helbig's Orientalist comments have motivated me to develop more convincing accounts of how tools and machines are being used in various corners of the world. Helbig's description of the Dayak goldminers shows us that "simple" tools may well be more appropriate and suitable than advanced, mechanical technologies, and some of his photographs remind us that modern technologies are volatile and transitory.[14] Helbig's nationalistic remark, about the Dayak using German products in the middle of the jungle, illustrates the fact that what he called modern and primitive technologies may very well coexist.

My approach has been affected in fundamental ways by postcolonial thinking and postcolonial history writing.[15] Such perspectives imply questioning and deconstructing dichotomies between "developed" and "underdeveloped" parts of the world, between center and periphery, between "the West and the rest," between the Occident and the Orient, between "advanced" and "primitive" cultures. Although many sources on which this book is based reflect an evolutionary view of human societies and their technologies, the microhistories in this volume do not reproduce such colonial paradigms. Rather than emphasize the impact of Western science and technology in various corners of the world, I recount histories of indigenous developments and cross-cultural encounters, and I analyze processes of hybridization and resistance.[16]

The eight chapters that follow are meant to counteract standard story lines in the global history of technology. The chapters reveal the complex

[12] Ibid., pp. 5, 51.

[13] Sardar, Ziauddin, *Postmodernism and the Other: The New Imperialism of Western Culture*. London: Pluto Press, 1998.

[14] Helbig, *Durchquerung*, p. 365.

[15] Ashcroft, Bill, Gareth Griffiths, and Helen Tiffin, eds, *The Post-Colonial Studies Reader*, 2nd ed. London: Routledge, 2006. Majumdar, Rochona, *Writing Postcolonial History*. London: Bloomsbury, 2010.

[16] Bhabha, Homi K., *The Location of Culture*. London: Routledge, 1994.

character of technological change and use. At the forefront is the idea that inventions do not simply emerge in a particular setting and spread gradually and uniformly across the globe. The common notion of one-way "technology transfer" only rarely describes accurately the relation between various nations or continents in an appropriate manner. Whereas people in one region may adopt innovations willingly, inhabitants in other regions may reject them outright. History teaches us that once-popular technologies can be later discarded.[17] As I will show in several chapters, new technological solutions and long-established technologies were often employed in synergy.[18] As such, I have de-emphasized the idea of new technological solutions that circulate globally; instead, I analyze local technological landscapes and material cultures.[19]

The book's main chapters are organized in rough chronological order. The first three chapters refer to the nineteenth century, a mainly colonial period; these chapters also trace developments to precolonial times. Chapters 5 and 6 focus on the interwar period and the first two decades after the Second World War. Chapters 7–9 are fully rooted in the postwar period. These chapters also examine postcolonial structures and the change they represented. In the concluding chapter, I touch on standard forms of periodization and their presumed usefulness in helping us to understand the global history of technology.

In contrast to the great majority of works in the history of technology, this book investigates the world beyond Europe and North America. By recounting histories from Asia, Africa, and Latin America, I hope to shed light on the rich and varied technological history of these continents. The microhistories in this book are also meant to incorporate recent insights by global historians, and tie local events to regional and global processes.[20] For example, in Chap. 4, I recount the global history of sugarcane cultivation and sugar manufacturing, while showing that the case of Northern India differs substantially from the more familiar histories, from Brazil, Cuba, and Java. Helbig's brief description of Gunung Mas is instructive in

[17] Perrin, Noel, *Giving Up the Gun: Japan's Reversion to the Sword, 1543–1879.* Boston, MA: David R. Godine, 1979.

[18] Edgerton, David, *The Shock of the Old: Technology and Global History since 1900.* London: Profile Books, 2006.

[19] Bray, Francesca, "Flows and Matrices, Landscapes and Cultures," *ICON: Journal of the International Committee for the History of Technology* 22, 2016: 8–19.

[20] Ghobrial, John-Paul, ed., "Global History and Microhistory," *Past & Present*, Supplement 14, 2019 (special issue).

that it illustrates the transient nature of Western influence and presence. In addition, it shows that Chinese goldminers and traders had for centuries been active in Kalimantan—the part of Borneo that belongs to today's Indonesia. The region had been an integral part of a trans-Asian gold trade network well before the Dutch arrived.[21]

Reclaiming the Term Technology

Standard definitions of the word technology pose a challenge for anyone interested in developing a global history of technology.[22] Today, "technology" is strongly associated with smartphones, computers, and other digital, high-tech solutions. In contrast to this connotation, this book embraces every form of technology, from the pejoratively named "low-tech" solutions to the often exalted "high-tech" ones. That means the metal mortar-and-pestle used by the Dayak people to crush rock fragments assumes the same status as the steam-driven grinding machine installed on Gold Mountain by the Dutch company Mijnbouw Maatschappij Kahajan. Indeed, we are obliged by history to use the word technology for any artifact or system and the knowledge and skill employed to use it.

Burong soho was one of the many technologies employed by gold workers in Kalimantan. According to geologist Theodor Posewitz, *burong soho* was a "special kind of bird" which was used as a kind of "dowsing instrument" in the search for gold deposits.[23] When the gold seekers approach an area rich in gold, the bird "begins to sing." Another technology employed was the "witchcraft basket"—as missionary Hugo Haffner somewhat contemptuously called it—which was placed at the entrance of a newly opened pit to "ask the spirits for their blessings."[24]

The technologies which Posewitz and Haffner called "a special kind of bird" and "witchcraft basket" were part of the Dayak goldworkers' toolkit, along with bamboo ladders and clay pots. Miners used spades to dig mines, crowbars to break rock into smaller pieces, and stone hammers

[21] Posewitz, Theodor, "Das Goldvorkommen in Borneo," *Mittheilungen aus dem Jahrbuche der kön. ungarischen geologischen Anstalt* 6, 1883: 175–190.

[22] Schatzberg, Eric, *Technology: Critical History of a Concept*. Chicago, IL: University of Chicago Press, 2018.

[23] Posewitz, "Goldvorkommen," p. 183.

[24] Haffner, Hugo, "Unsere Dajak auf dem Goldfeld bei Tewah," *manuscript*, 1937, in: Basel Mission Archives/Mission 21, Basel, Switzerland: File No. B-11.04.

and anvils to pulverize those pieces.[25] The locals had longstanding experience in using *dulang-dulang* (wooden pans) for gold-panning. According to rumors of the time, some parts of Borneo were so rich in gold that people could use sticks soaked in resin to retrieve gold nuggets from the ground.[26] By contrast, the Chinese miners who in the eighteenth century traveled to Borneo in large numbers created a technical system of excavation sites, dams, waterwheels, and washing channels to exploit deposits close to the surface.[27] In turn, gold-seekers from Europe brought petroleum lamps, steam engines, and locomotives. When Helbig visited Gunung Mas, several—though not all—of these technologies coexisted. Together, these tools and practices comprised Borneo's technological landscape of gold extraction.

I argue that the objects which Europeans labeled "a special kind of bird" and "witchcraft basket" were integral parts of Dayak material culture. Importantly, this culture was dynamic, and, progressively, miners integrated German mallets, chisels, and storm lanterns, for example. Despite their acceptance of these new implements, the Dayak people retained their belief in what Helbig called the traditional "legend" of how the gold had come to Borneo in the first place. The presence of these coexisting technologies and beliefs render it impossible to draw a clear line between the material and immaterial aspects of Dayak culture.

I suggest in this book that we can come to a better understanding of technology if we define it in terms of "material culture."[28] According to Anne Gerritsen and Giorgio Riello, historians of the Early Modern Period, a history of material culture focuses on objects "in the everyday practices that shaped past lives."[29] Although the following chapters describe material cultures in selected regions of Africa, Asia, and Latin America, I recommend we apply the concept of "material culture" to every other region

[25] Posewitz, Theodor, *Borneo: Entdeckungsreisen und Untersuchungen. Gegenwärtiger Stand der geologischen Kenntnisse. Verbreitung der nutzbaren Mineralien*. Berlin: R. Friedländer & Sohn, 1889, pp. 264–265.

[26] Dewall, H. von, "Aanteekeningen omtrent de noordoostkust van Borneo," *Tijdshrift voor indische Taal-, Land- en Volkenkunde* 4, 1855: 423–458.

[27] Jackson, James C., *Chinese in the West Borneo Goldfields: A study in Cultural Geography*. Hull: University of Hull Publications, 1970.

[28] Cf. El Hariry, Shorouk, et al., "Toward a Global History of Material Culture," *Technikgeschichte* 88, 2021: 178–182.

[29] Gerritsen, Anne, and Giorgio Riello, "Introduction: Writing Material Culture History," in: idem, eds, *Writing Material Culture History*. London and New York: Bloomsbury Academic, 2015: 1–13, here: p. 4.

of the world—including Europe and North America. For the historian of technology, applying a material-culture approach implies treating heavily industrialized and less industrialized regions with the same methods—in an unbiased manner.

The microhistory method, too, is universally applicable. From the 1960s onward, when "microhistory" initially developed as a historiographic approach, it was employed almost exclusively by historians of Europe.[30] In sharp contrast to traditional topics like kings, churches, and wars, microhistorians developed an interest in the worldviews and daily lives of ordinary people: *la vie privée* in French and *Alltagsgeschichte* in German.[31] Of paradigmatic importance to the development of *microstoria*—as this approach is called in Italian—is Carlo Ginzburg's analysis of the ideas and convictions of a sixteenth-century North Italian miller.[32] To a large extent microhistory became associated with a cultural approach. For example, German historian Hans Medick applied cultural-historical methods in his impressive work about a small linen-weaving community in southwest Germany.

Microhistories of Technology can be read as an experiment in applying the microhistory and cultural-history approach in Asia, Africa, and Latin America. Methodologically, this book was inspired by Medick's observation that microhistorians unravel "less spectacular events and actions by telling small stories of daily life," and zoom in on the "objects of material culture."[33] As Medick himself shows, however, such an approach does not preclude the investigation of connections between local processes and processes that take place on national—and even international—levels. Medick's book shows that despite substantial structural changes that took place from the eighteenth through the early twentieth centuries, the linen weavers of Laichingen in Württemberg, Germany, stalwartly provided customers in France and Italy with high-quality products. This book does not follow a single community for a century and a half, though it does contain stories about the use of technology in everyday settings from various communities—from the mid-nineteenth to the late twentieth century.

[30] Schulze, Winfried, ed., *Sozialgeschichte, Alltagsgeschichte, Mikro-Historie: Eine Diskussion.* Göttingen: Vandenhoeck & Ruprecht, 1994.

[31] Ghobrial, John-Paul A., "Introduction: Seeing the World like a Microhistorian," *Past & Present*, Supplement 14, 2019 (special issue): 1–22.

[32] Ginzburg, Carlo, *The Cheese and the Worms: The Cosmos of a Sixteenth-century Miller.* Baltimore, MD: Johns Hopkins University Press, 1980 (orig. 1976).

[33] Medick, Hans, *Weben und Überleben in Laichingen 1650–1900: Lokalgeschichte als Allgemeine Geschichte.* Göttingen: Vandenhoeck & Ruprecht, 1996, p. 21.

Protagonists and Sources

Often, material objects outlive human beings. That said, we cannot imagine cultures without people. The protagonists of the following chapters range from Peruvian cooks to Tanzanian electricians, West African kola-nut traders to Indian rural laborers. Given that written historical sources do not always name the various actors, some of the protagonists in these microhistories have been lost to anonymity. Whenever possible, however, I have tried to supplement the archival sources with interviews; for ethical reasons, the interviewees' identities in this research must also remain anonymous.

Most of the microhistories presented in this book rely on sources—including oral sources—provided by doctoral candidates and postdoctoral researchers with whom I have collaborated. Their research experience spans Asia, Africa, South America, and Europe. As detailed in the acknowledgments at the beginning of the book, these colleagues have located and retrieved unpublished and published sources from archives and libraries across the globe. In several cases, the researchers carried out in-person interviews. I gathered material from East African, Western European, and Indian archives and libraries. This book is based on our unique collection of primary-source material, which gives readers insight into the daily practices, tools, and machinery of ordinary people in various cultures.

In all cases, the focus is on people who led "ordinary"—rather than privileged—lives. Inspired by Ginzburg's *microstoria* and Medick's *Alltagsgeschichte*, I argue that historians of technology have a great deal to learn from the way mundane tools are used in everyday settings. Featured here are people who built their own homes with adobe bricks; washed and folded their own menstruation pads; and earned a living brewing beer and making sugar for their communities. And while engineers, scientists, and architects populate these narratives, they are purposely placed on the sidelines. Rather than being set in high-tech laboratories or modernist urban neighborhoods, the stories in this book take place in missionary stations, tropical rainforests, and working-class kitchens, for example.

Anyone who tries to write the history of ordinary people and daily life must contend with the limitations of archives as repositories of research material. National archives tend to be biased toward official documents produced by state institutions; archives seldom contain the writings of ordinary people—making it difficult to write what E.P. Thompson, in his seminal history of the English working class, called "history from below."[34] Historians who work with archival material in former colonial settings must grapple with another grave problem: only rarely do historians of colonial periods find primary sources produced by members of indigenous populations. The great majority of sources stem from the colonial rulers, and the material reflects the ideas and activities of the local population only indirectly—if at all. The following chapters have been written with these limitations in mind. In addition to using interview material, I refer to written documents and images from travel accounts and diaries, magazines and newspapers, for example. Some of this material comes from individuals' private papers as well as the archives of various nongovernmental organizations.

Each story in this book is based on a carefully selected set of historical sources. Taken together, these microhistories show how differently people approached, developed, and used technology in distinct cultural settings. These microhistories are meant to problematize the notion that the standard global history of technology is one sided and linear, a narrative of growing uniformity and homogeneity. And, as mentioned, while "globalization" and "mechanization" have been important concepts, neither capture the complexities of history. In writing these microhistories, I followed anthropologist Ulf Hannerz's twofold advice that "globalization has to be brought down to earth," and that "the local has to be brought up to the surface."[35] The people described in this book are not isolated from the wider world, nor are their lives determined by this world.

Further, the chapters make fundamental theoretical contributions to our understanding of the character of technology; to the relationship between materiality and culture; and to the global history of technology. For example, in Chap. 4, I use the microhistory of sugar manufacture in Northern India to argue that we need to redefine the well-known concept of "appropriate technology." In Chap. 7, I suggest that concepts such as "slum" and "informal settlement" to describe Nairobi's low-income areas may be better characterized as "flexible settlement." And in Chap. 8, I address the question to what extent the concept of "Americanization"—so often used by historians in reference to post–Second World War

[34] Thompson, Edward P., *The Making of the English Working Class*. London: Gollancz, 1963.
[35] Hannerz, Ulf, *Transnational Connections: Culture, People, Places*. London: Routledge, 1996, pp. 19, 28.

Europe—is helpful in describing the proliferation of consumer goods in South Korea.

This book is not a synthetic work with universal ambitions. Instead, I apply a microhistory approach to provide iconic examples from the global history of technology, from the mid-nineteenth century onwards. Some stories exemplify the innovative power of human beings, while others illustrate the usefulness of hybrid, cross-cultural solutions. Indeed, these stories are meant to expand the historian's geographical horizon; they are also meant to show how the small worlds of local actors are connected to other parts of the globe.

PART I

Nineteenth-Century Ways of Life

CHAPTER 2

Building Missionary Stations in Southeast Asia: Nias Islanders Deploy Adzes

> Early at 10 am. I arrived from Sifaoroasi with 6 pupils, tied up Hektor [to a tree] and gave him some grass, and began to work. We cleared the undergrowth and prepared a place to sleep, i.e., an outback hut. This building was also meant to house the two carpenters from Sifaoroasi. Something like this is only possible in the tropics, where you find everything you need close at hand. You turn a couple of trees into poles, and vines are much easier to handle than expensive European cords or nails. Bamboo shrubs from the area serve as roof laths, and leaves from sago-palm trees are used as thatching. Wood from the forest makes a primitive floor. At 5 pm. everything was finished. The horse was eating its bundle of grass, water was boiling in the pot above the fire, and the hammock was already hung. We had rice with salt and onion. Served on a banana leaf, it tasted better than any delicacies imaginable. The boys worked with joy. To accomplish something in the fresh air made them just as happy as me. All of us had honestly earned our supper.[1]

This quotation is taken from the published diary of Eduard Fries, German missionary to the Dutch East Indies (today's Indonesia). In the entry from 1907, Fries goes on to describe how, after supper, the party enjoyed the black silhouette of the distant mountains. When the fog began to cover the valley, the group gathered inside the simple shack to sing

[1] Fries, Eduard, *Rundbriefe* 4, 1907, Berlin: Gutenberg, Circular No. 31, p. 37.

M. Hård, *Microhistories of Technology*, Palgrave Studies in the History of Science and Technology,
https://doi.org/10.1007/978-3-031-22813-1_2

"Christian songs." Fries speculates that the party's loud voices could be heard throughout the nearby village of Siwalubanua.[2]

This rhapsodic narrative contains many of the hallmarks of romantic Western travel accounts: the landscape replete with natural resources, the naturalness of indigenous techniques, the almost mythically flavorful local cuisine, the afterglow of a hard day's work. Read in a contemporary context, the quote may remind us of, say, a blog post by a writer describing a trendy adventure-vacation or a survival-training course. According to this kind of narrative, the Western adventurer journeys to an "exotic" place; does manual labor in the unfamiliar environment; completes a task using only on-site materials; and, when the experience is over, takes pride in the accomplishments and returns home, "safe and sound."

Fries was neither a tourist nor a recreational survivalist; he was a Protestant missionary to Nias, a small island off the western coast of Sumatra. Like other Christian missionaries of his time, Fries' task was to proselytize and to convert the local population and bring what he and his colleagues called "cultural progress" to "semi-barbaric" peoples.[3]

When Fries embarked on his trip to Siwalubanua, he was thirty years old. The purpose was to establish a branch office to Sifaoroasi, where he was stationed. After months of negotiations with local chiefs, Fries had been granted permission to build a school and a teacher's home in Siwalubanua. More precisely, Fries was authorized to hire a group of local craftspeople to erect these buildings. The Nias carpenters, who were meant to stay in the simple hut during construction, did not disappoint Fries: they quickly completed the first building. Soon after, the teacher, a Nias convert, began to gather pupils from Siwalubanua and the surrounding villages and homesteads in the southeast corner of the island.[4]

Toward the end of 1903, the Rhenish Missionary Society, headquartered in Barmen, Prussia, had dispatched Fries to the Dutch East Indies. Fries' journey via the Suez Canal and across the Indian Ocean took place without incident, and his ship docked in the region he called "India," a couple of days ahead of schedule. After a brief stay on Sumatra, Fries arrived on Nias in January, 1904. He would remain on the island for a

[2] The majority of the archival material for this chapter was researched and reported by Dr. Mai Lin Tjoa-Bonatz.

[3] Angerler, Hans, "Mission, Kolonialismus und Missionierte: Über die deutsche Batakmission in Sumatra," *Beiträge zur historischen Sozialkunde* 23 (2), 1993: 53–61, here: 56.

[4] Fries, *Rundbriefe 4*, 1907, Circular No. 31, pp. 33–38.

decade-and-a-half. From the time of his arrival until the outbreak of the First World War, Fries communicated with the Rhenish Missionary Society, back home, on a regular basis. More than two-hundred members of the Society subscribed to Fries' so-called circulars (*Rundbriefe*), a mix of diary entries, observations, and personal reflections. Eduard's brother, Wilhelm Fries, edited the circulars and made sure they were printed and collected in annual volumes. During Fries' first ten years on Nias, he submitted a total of sixty-five circulars.

Fries and his cohort used circulars as a communications tool. The elevated *Rundbriefe* elaborated on the "civilizing missions" in order to mobilize financial support for the Society. More than a century later, historians can use the information in the *Rundbriefe* to glean insights into the daily lives of people—as well as to discover the microhistory of life at missionary stations.[5]

Fries was not the first Rhenish missionary in the area. Eight years after its founding in 1828, the Rhenish Missionary Society had established its first station in the Dutch East Indies on the island of Borneo.[6] One of the pioneers, Ludwig Nommensen, had lived on Sumatra and Nias for more than forty years before Fries arrived. One of Nommensen's colleagues, Ernst Ludwig Denninger, had traveled from Sumatra to Nias in 1865, to set up the first station there. Accordingly, when Fries arrived on Nias, he benefited from an existing network of missionary stations. This network included a series of global infrastructures and the support of the Dutch colonial administration.

When Fries made landfall in Gunungsitoli—originally a military station, and a small settlement as of 1900—on the east coast of Nias, he was welcomed by other members of the Rhenish Mission. A small pile of letters from home awaited him in his temporary dwelling at the Ombolata missionary station, south of Gunungsitoli. Anyone who wished to contact

[5] A complete set of Fries' circulars are available at the Archive of the Rhenish Mission, housed in the Library and Written Archives of the Museum Foundation of the United Protestant Mission (Archiv- und Museumsstiftung der Vereinten Evangelischen Mission, hereafter: VEM) in Wuppertal-Barmen, Germany. For further information on Fries' life, see Humburg, Martin, Dominik Bonatz, and Claus Veltmann, eds, *Im "Land der Menschen": Der Missionar und Maler Eduard Fries und die Insel Nias*. Bielefeld: Regio, 2003.

[6] Menzel, Gustav, *Die Rheinische Mission: Aus 150 Jahren Missionsgeschichte*. Wuppertal: Verlag der Vereinigten Evangelischen Mission, 1978.

Fries needed no zip code; "Missionar Fries, NIAS, Sumatras Westkust, via Genua, Ombolata. Nederlandsck Indie," was enough.[7]

Fries' journey from Genoa to the Dutch East Indies had filled him with excitement. In his diary he explains that he could watch the "infinite, eternal sea" for hours and days on end; distant shorelines fueled his imagination. During a visit to the lower deck of the ship, Fries marveled at the colors and smells of the "mosaic" in the third- and fourth-class decks: "Chinese with long pigtails … Arab pilgrims to Mekka … beautiful Indians, sweet Malays". Fries watched "the hundreds of coolies and Chinese … the slaves of the whites" who were "dragging coal" in the harbors. While he wrote glowingly, if naïvely, of the human "mosaic" aboard the Dutch steamer, Fries was apparently relieved to be able to maintain some distance from servants and "tricksters" who allegedly tried to steal the first-class travelers' money.[8]

A romantic, exotic tone permeates Fries' account. His description of himself as both involved and distanced reminds us of what sociologist John Urry calls "the tourist gaze."[9] Indeed, despite having embarked on the trip to Indonesia for the purposes of carrying out a religious mission, Fries viewed his surroundings with eyes similar to those of a tourist. It was only when Fries began to live near—and work with—the Niasans that he began to interact directly with people of other backgrounds.

The operative question I explore in this chapter is: What happened when Fries and other missionaries took up residence in distant lands for longer periods? What daily routines did they develop in their new environments, and how did the indigenous people on Nias and Sumatra respond to the missionaries' presence? What practical skills did the missionaries bring with them, and what knowledge did they acquire from the locals?

In exploring this question of cultural interaction, I analyze direct encounters between Europeans and members of the indigenous population. The physical sites where different groups met and interacted are of equal interest. Missionary stations were iconic places where Europeans and locals observed and influenced each other. Like the local marketplace, where the Protestant male Rhenish missionaries, their wives, and servants

[7] Fries, *Rundbriefe 2*, 1904, Circular No. 5, p. 5.

[8] The quotes in this paragraph are from Fries, *Rundbriefe 1*, 1903, Circulars No. 1, 2, and 3.

[9] Urry, John, *The Tourist Gaze: Leisure and Travel in Contemporary Societies*. London: Sage, 1990.

would have purchased produce from local farmers, the missionary station can be called a "trading zone."[10]

This figurative interpretation of a trading zone acknowledges implicitly that missionaries and locals did not trade in material goods alone; they also bartered technical knowledge and skill. When building the small hut in Siwalubanua, the Nias carpenters—as well as Fries—used materials from their immediate surroundings and applied regional techniques. Fries did not merely watch the local craftspeople; he worked alongside them, using skills and knowledge he had acquired when his main missionary station had been built in Sifaoroasi, two years earlier. The idea of using palm leaves to cover a roof was, by then, no longer new to him.

In describing the "trading" of objects and knowledge between groups on Nias in the late nineteenth and early twentieth centuries, I borrow the trading zone concept from Peter Galison's work in the history and philosophy of modern science.[11] Galison repurposed "trading zone," applying the term to a microphysics lab in the late twentieth century. In analyzing the social and cognitive relations in the science lab, Galison overturns the idea that scientists necessarily take the lead—and that engineers merely apply the ideas of science. No such hierarchy exists in practice, according to Galison. On the contrary, to be able to carry out viable lab experiments, scientists, engineers, and technicians must "trade" information and learn to collaborate—despite their inevitably differing epistemologies and perceptions of the world.

The missionary station, as a workplace, may seem far afield from the microphysics laboratory of Galison's research. However, both sites host interactions between individuals with different education, experience, and socialization. Similarly, the construction and management of missionary

[10] For a more elaborate discussion of the concept "trading zone" in this connection, see Hård, Mikael, and Mai Lin Tjoa-Bonatz, "Trading Zones in a Colony: Transcultural Techniques at Missionary Stations in the Dutch East Indies, 1860–1940," *Social Studies of Science* 50 (6), 2020: 932–955. Not the Rhenish Society, but a few other societies dispatched women as missionaries overseas. Catholic nuns also carried out missionary activities; cf. Stornig, Katharina, *Sisters Crossing Boundaries: German Missionary Nuns in Colonial Togo and New Guinea, 1897–1960*. Göttingen: Vandenhoeck and Ruprecht, 2013.

[11] Galison, Peter, *Image and Logic: A Material Culture of Microphysics*. Chicago: The University of Chicago Press, 1997. More recently, Pamela O. Long applied the concept in the article "Trading Zones in Early Modern Europe," *Isis* 106 (4), 2015: 840–847. Originally, the concept comes from anthropologists concerned with the development of hybrid languages in former colonial settings; see, e.g., Romaine, Suzanne, *Pidgin and Creole Languages*. London: Longman, 1988.

stations required some form of exchange between Westerners and locals. To characterize the encounters between Christian missionaries and the people they tried to convert, historian Jürgen Osterhammel uses the apt term "symmetrical interaction." Although missionaries and colonial powers tried to impose their worldview on indigenous peoples, day-to-day life at missionary stations was to a large extent formed by "local conditions."[12]

Indeed, what distinguishes the missionary station as a trading zone is the fact that information traveled both ways. Although communication between Europeans and Asians was imperfect, that information—as well as material goods—was often exchanged successfully. For example, to build houses that appealed to European taste, local carpenters "received" information about—and adopted—technologies that were new to them. Carpenters learned to use tools and materials that Fries had brought with him from Germany. Some builders were so eager to adopt Western tools that Fries felt compelled to end his short account of constructing the Siwalubanua hut with the following coda:

> It is, by the way, interesting to observe how the indigenous long to adopt our cultural achievements. Unfortunately, they do not realize that we tend to prefer the freely available and often more durable indigenous products. And, as a result, they forget their own, often more practical techniques.[13]

By twenty-first-century standards, the tone of Fries' reflections can be considered paternalistic and Orientalist. Nevertheless, his reportage shows that some indigenous craftspeople on Nias readily appropriated foreign technologies to serve their own needs and interests. In the house-building "trading zone" of Nias, the Batak people learned to join wood with iron nails as surely as Fries learned to thatch a roof with local palm leaves.

Pioneers and "Martyrs"

British and U.S. missionary societies had long been active in the area before Fries arrived on the scene. Even Nommensen had predecessors: in 1820, the Baptist Missionary Society of Britain had dispatched to Sumatra

[12] Osterhammel, Jürgen, *The Transformation of the World*. Princeton: Princeton University Press, 2014 (orig. 2009), pp. 891, 893.

[13] Fries, Eduard, *Rundbriefe 4*, 1907, Circular No. 31, p. 37.

Richard Burton and Nathaniel Ward.[14] In an account of that trip, the two missionaries express deep admiration for the Toba Batak—a people living in the western parts and the highlands of northern Sumatra—and their building accomplishments:

> We were scarcely less interested by the internal appearance of the villages, than we had previously been by that of the surrounding country. The one subordinate to our host consisted of twenty-four houses, in a straight line, with the gable ends uniformly facing the street ... The houses were constructed generally of excellent materials, exhibiting marks of superior workmanship, and in many instances ornamented with carving and paint. The villages were clean; and the females occupied in the manufacture of cloth, and surrounded by numbers of playful children, afforded a pleasing idea of industry, health, and domesticity.[15]

Unsurprisingly, Burton and Ward's interpretation of Batak society is shot through with Western ideals of uniformity, cleanliness, and industriousness. Despite its Western bias, their observations are important, however. The earliest missionaries were impressed by how the local people built their houses, a sentiment echoed by Fries some eighty years later.

Apparently, the fascination was mutual. Burton and Ward reference the Toba Batak people's interest in the White men and their belongings. During an expedition to the Sumatran highlands in 1824, Burton and Ward were encircled by villagers, who had seldom or never seen Europeans. Equally intriguing were the British missionaries' objects: a "double-barrelled gun, ... a telescope, a mariner's compass, a pair of spectacles, a case of mathematical instruments, some printed books."[16] No mention is made of whether or not the locals were allowed to handle any of the objects.

A decade later, the young Baptists Henry Lyman and Samuel Munson, from the United States, undertook a journey to Sumatra and Nias to explore the possibility of setting up a missionary station in the area. In his

[14] Cox, Francis Augustus, *History of the Baptist Missionary Society, from 1792 to 1842, Vol. I*. London: T. Ward and G. & J. Dyer, 1842, p. 353.

[15] Burton, Richard, and Nathaniel Ward, "Into Toba Batak Country, 1824," in: Reid, Anthony, ed., *Witnesses to Sumatra: A Travellers' Anthology*. Oxford: Oxford University Press, 1995: 175–192, here: pp. 177–178.

[16] Ibid., pp. 176–177.

diary, Lyman offered advice for future missionaries. Settlers had to make meticulous preparations:

> If a missionary wishes to reside in the island, he would do well … to bring with him his household furniture, iron work for building, and stores, such as sugar, coffee, tea, etc., etc., and goods for purchasing the necessaries of life, and making some few presents. Goods should consist of tobacco, iron and steel, and coarse cloths particularly. When arrived here, he should pay his respects immediately to all the chiefs in the district … He should ask them for a piece of land, or, rather, select a piece and tell them he wishes to build upon it. His timber he will purchase cheap. Laborers he will obtain, who will make his house after a fashion, if he can have patience to give them an exact plan, and show them all the parts. … He would do well to raise his own vegetables, fruits, fowls, hogs, sheep, and to keep a horse, with a saddle and bridle. Rice and potatoes he could always purchase cheap, with goods.[17]

This excerpt highlights several important realities. Perhaps predictably, Lyman recommends that missionaries import materials to create a decent home; he also advises missionaries to bring ceremonial gifts for chiefs and neighbors. Less predictable is Lyman's statement that the missionaries could rely on local craftspeople to construct the necessary buildings. Batak carpenters had the skills to build houses that were in line with missionaries' wishes. While indigenous people may not have worked previously with iron building materials, Lyman was certain they would understand how to do so.

The excerpt also establishes that missionaries could expect more than a subsistence way of life, given that some produce was locally available. Passages from Lyman's other records and letters show that people on Sumatra and Nias traded coconut oil, sago starch, and edible sea slugs, "so much prized by the Chinese." People also manufactured textiles from cotton and silk, and some islanders were trained blacksmiths and goldsmiths.[18]

Before Lyman and Munson could establish themselves as missionaries, they—together with one of their servants—were killed by a group of Batak warriors who objected to White men entering their country. This tragic ending later motivated Lyman's anonymous biographer and editor of his

[17] *The Martyr of Sumatra: A Memoir of Henry Lyman*. New York: Robert Carter & Brothers, 1856, pp. 380–381.

[18] *Martyr*, pp. 330, 354, 356.

diary to give him the epithet "the martyr of Sumatra."[19] It would take another half-century until missionaries could establish safe outposts in remote areas of Sumatra and Nias. Only after two bloody wars around 1880 did Dutch military forces seize authority in the region—although pockets of resistance remained into the early-twentieth century. This so-called pacification of the local population enabled missionary organizations in Europe and North America to establish new stations. Decisions about where to expand their activities were made in consultation with the missionaries already on the ground, and they required permission from the local Dutch authorities. Governors usually allowed missions to expand their activities into "independent districts" only when the missionaries' safety could be guaranteed.[20]

The missionaries' period of residence in faraway countries was characterized by a certain degree of ambivalence. On the one hand, they were eager to learn the local language, and they went out of their way to establish close contacts with the indigenous population. Before the Rhenish Society sent their missionaries overseas, it offered language courses to make sure the missionaries were able to communicate with potential converts. On the other hand, once missionaries found themselves on-site, they made an effort to maintain relative physical distance from the local population and their ways of life.

After being assigned a region, the missionary initially had no choice but to rent a dwelling from one of the locals. This was the case for Nommensen, as a young missionary. In 1863, when he first arrived in Baros, "a destitute place" on the Sumatran coast, he moved into a house owned by a Chinese villager. According to Nommensen's account, the dwelling "was in such a bad state that it kept neither wind nor rain out." After some negotiation, the Chinese landlord agreed to reduce the rent if Nommensen agreed to renovate the house at his own expense.[21]

Daily life in simple houses proved to be distinctly unromantic. Indeed, complaints about the limited space and unsanitary conditions in ordinary homes is a common trope in missionary narratives. Women missionaries to China—who were allowed to proselytize despite not being ordained

[19] *Martyr.*

[20] *Berichte der Rheinischen Missions-Gesellschaft* (hereafter: *BRMG*) ["Reports of the Rhenish Missionary Society"], 1862, p. 78.

[21] *BRMG*, 1863, p. 131.

priests—routinely reported on their dark, uncomfortable houses, as well as the armies of bedbugs they battled at night.[22]

The Rhenish emissaries to the Dutch East Indies recounted similar experiences. "Brother" Schrey, stationed in Sibolga, on Sumatra's west coast, vented his displeasure:

> A whole house had been prepared and cleaned for us. Still, it was so dirty that we got covered in black, and the ceiling was so low that we were constantly bumping our heads. But the worst were the countless bugs that visited us during the night….[23]

One of Schrey's colleagues was equally unhappy about having to live in a dwelling that looked like "a mediocre hen or duck house."[24]

Often, concerns centered on soot from indoor fireplaces. Given that the houses lacked chimneys, missionaries had difficulty breathing. However, they soon realized the fireplace's key advantage: it kept mosquitos out of the buildings. One missionary claimed he would rather be "smoked" inside his own house than devoured by aggressive insects.[25] Missionaries also objected to the fact that walls and floors were not entirely solid. Made of wooden planks or bamboo branches, houses featured narrow slits between the wood and the twigs. These openings provided ventilation as well as vulnerability to the elements. It also meant that noise and odors found their way into missionaries' living spaces. Historical sources are rife with the laments of Europeans, who cite the stench of animals and unwashed people and the lack of privacy. Some of the grievances are plausible: traditional houses on Sumatra were elevated on stilts; the space beneath was used as a garbage dump, privy, and pigpen.

When missionary Jonas Klammer and his wife arrived in Sipirok, North Sumatra, in 1861, they were more content. In the Sumatran highlands they were able to buy a new house that served their needs. In a letter to her family back in Germany, "Sister Klammer," as she is called in the Society's reports, describes the layout and the construction technique.

[22] Freytag, Mirjam, *Frauenmission in China: Die interkulturelle und pädagogische Bedeutung der Missionarinnen untersucht anhand ihrer Berichte von 1900 bis 1930*. Münster: Waxmann, 1994, p. 116.

[23] *BRMG*, 1882, p. 213. In the available sources, missionaries' first names are seldom explicitly mentioned.

[24] *BRMG*, 1862, p. 65.

[25] *BRMG*, 1862, p. 65.

The walls were made of tree bark, and the roof was covered with "*alang-alang*, a kind of grass." The living room and four other rooms were separated by bamboo walls, approximately two-and-a-half meters high. Only the living room was covered by a ceiling per se; hewing to the local building style, the other living spaces were covered only by the steep, grass-thatched roof. This design allowed the air to circulate effectively, though it carried the disadvantage "that you do not only hear every word being said in the neighboring room; you can even hear people breathing." An adjacent building, with "kitchen, horse stable, and a study," was yet to be erected on the premises. Since the main building still lacked doors and window shutters, Jonas Klammer himself was obliged to install them.[26]

To survive, missionaries had to take part in both local and global commodity markets. In contrast to what people at home apparently believed, Sister Klammer claims that, on average, living expenses on Sumatra were higher than in Germany. In Sipirok—approximately 1,000 meters above sea level and several days' walk from the coast—traders at the local market (*passer*) were able to demand four times more for their produce than for the fruits and vegetables sold in the larger coastal villages. Considering the butter Sister Klammer bought for her family had been shipped to Sumatra from the Netherlands, her cost-of-living comparison may well have been correct.

The further inland they resided, the higher the missionaries' cost of living climbed. Living far away from ports and population centers also meant enduring what the emissaries considered to be primitive ways of life. After the turn of the century, for example, the Rhenish missionary Brother Guillaume had arranged to have a house built for himself close to Lake Toba, in the middle of Northern Sumatra. The house did not, however, fully please him: the floors were made of bamboo, the walls of plaited reed (*arong*), and the roof was covered with "long, dry grass." Guillaume grumbled, "...everything is woven together by means of rattan; the floor moves after each step, and the mountain wind blows through the whole house at liberty."[27]

Once they were established in their assigned villages, missionaries could exercise more influence on the design of their homes and yards. For example, Wilhelm Thomas, who moved into his newly built house in Ombolata, on Nias, in 1874 (thirty years before Fries), writes proudly of his

[26] The quotes in this paragraph are taken from *BRMG*, 1862, pp. 74–79.

[27] The quotes in this paragraph are taken from *BRMG*, 1907, p. 95.

"soothing oasis in the wilderness." Thomas had built a relatively small (thirteen-by-six meter) house, which nonetheless boasted a "large drawing room," a bedroom, a storage cabin, a study, and a "room for visitors with two camp beds." In line with local building practices, the house was set on stilts and featured a large porch. In contrast with local practices, however, the rooms were painted white, and the structure's outer walls were whitewashed. The choice of white (achieved by using lime) likely reflects Thomas' wish to differentiate his own estate from the darkness and dirt that he and his cohort associated with indigenous dwellings. The fruit trees and flowers in Thomas' garden were meant to underscore the difference between Western perceptions of hominess and the surrounding tropical rain forest.[28]

Building Styles Intersect

These descriptions illustrate the hybrid nature of the missionaries' homes.[29] On the one hand, the goal was to design these stations according to European ideals of hygiene and spatial order. Instead of following the local Niasan and Batak tradition of allocating the interior to one large, multifunctional room, missionaries usually requested that their quarters be divided into several single-function spaces. Larger stations typically featured a bedroom, living room, nursery, dining room, study, and pantry, as well as guest rooms and even a storage cabin for medicine. Next to these quarters was a privy and a separate building with facilities for cooking, washing clothes, bathing, housing domestic animals, and storing equipment.[30] Missionary-station yards often included flowers and fruit trees, as well as a garden with various vegetables, most of which were European.

On the other hand, when it came to structural solutions, missionaries usually adopted local customs, materials, and techniques. Most dwellings were raised on stilts. Roofs were high and steeply pitched. This design allowed tropical rains to run off quickly, and it helped interiors to remain

[28] The quotes by Thomas in this paragraph are taken from Töpperwien, Annemarie, *"Seine Gehülfin": Wirken und Bewährung deutscher Missionarsfrauen in Indonesien 1865–1930.* Cologne: Köppe, 2002, pp. 47–48.

[29] Concerning "hybrid" solutions in the history of technology, see Hård, Mikael, and Andrew Jamison, *Hubris and Hybrids: A Cultural History of Technology and Science.* Cambridge, MA: MIT Press, 2005.

[30] Cf. floor plans drawn by missionary Arnold Momeyer in 1919; to be found at VEM: Archival No. 2.754 ("Gunung Sitoli").

reasonably cool. Buildings in the area featured an intricate timber framework, designed to combine sturdiness with flexibility—in part to prevent serious earthquake damage. To help residents contend with the region's tropical climate, the prototypic house had a large porch (*ember*), where much of daily life took place.[31] Bamboo was favored for building Niasan and Batak homes; the use of bamboo was clearly not a European tradition.

The hybrid nature of the stations reflected intense collaboration, during the construction phase, between indigenous craftspeople and missionaries. The missionary drew floor plans, determined the layout of the station, and oversaw construction. Often, they took an active role in the building activities. In preparation, missionaries brought with them European tools and construction materials: saws, hammers, nails, and other items made of iron. The religious emissaries sought to teach local carpenters how to use these imported tools and materials.

Despite the application of certain Western tools, missionaries remained dependent on local technologies. Successfully completing a missionary outpost required access to two categories of resources: construction materials and building expertise. Missionaries needed raw materials from the immediate environment: timber, reed, bamboo. Building large stations could take place only with the knowledge and skill that indigenous carpenters and blacksmiths contributed. During the construction phase, the missionary station became a trading zone, where Europeans and Asians bartered tools and other goods—as well as knowledge and skills.

Once the missionary had acquired a site for a new station and received permission from the Dutch authorities, the local chief, and the Rhenish Society, construction could commence. Securing enough building material and engaging experienced craftspeople was not always easy. One might assume that being situated in the rain forest would ensure an abundant wood supply, but this was not always the case. The widespread use of slash-and-burn cultivation tended to limit access to timber. In the mid-1880s, one missionary reported home that it had been "particularly difficult to get ahold of wood. In addition, people hardly understand how to make planks." It becomes clear from this and other reports from Sumatra that missionaries could not rely on untrained villagers; it was essential to employ skilled workers:

[31] For building methods in the area, see Waterson, Roxana, *The Living House: An Anthropology of Architecture in South-East Asia*. Singapore: Oxford University Press, 1991.

> The people in Si Gumpar … have brought together considerable amounts of wood for the building. Unfortunately, it cannot really be used, since they do not know how to handle it. Construction will thus require substantial amounts of money for carpenters' wages.[32]

The Rhenish Society provided its emissaries with funds for materials and labor. These funds were not always sufficient, and missionaries were sometimes forced to dig deep into their own pockets. In 1908, the administrative policies at the society's headquarters in Barmen changed. From that point on, missionaries were allowed to submit applications for building funds that reflected the projected building costs. Indirectly, the new policy yielded an advantage for historians: the ability to reconstruct building plans in greater detail.

A trove of building detail can be found in Arnold Momeyer's 1919 funding application to the Rhenish Society's Building Committee for a residential house and annex in Gunungsitoli, Nias. Intended to contain twelve rooms, the house was an elaborate undertaking. Momeyer's calculations underscore this: the wooden frame alone required 340 meters of square beams, twelve-to-fourteen centimeters thick. An unspecified number of wooden planks were needed for the walls and floors, along with 200 stilts on which the house would sit. The purpose was to design the building according to standard Niasan methods, including a sturdy truss construction. Rather than using *alang-alang* grass, reed, or corrugated iron ("not recommended for G. Sitoli"), Momeyer suggested covering the roof with "somewhat expensive" shingles. He also calculated the costs of bricks and cement, as well as "locks, latches, pendants, ironworks, etc." Momeyer estimated labor costs for carpentry, cement casting, and the digging of a well.[33]

In his application, Momeyer went out of his way to justify the cost estimates, apparently to ensure that the committee would approve his request for the 5,000 guilders. Sources indicate that Momeyer was not alone in making an expansive request for funds: the new policies of 1908 fostered a certain covetousness on the part of some missionaries. In a letter to the director of the building committee, a missionary named Ed. Wagner complains about this lax attitude toward building costs on part of his colleagues:

[32] Quotes in this paragraph come from *BRMG*, 1886, pp. 109, 114.

[33] The information and the quotations in this paragraph are taken from VEM: Archival No. 2.754 ("Gunung Sitoli").

> Earlier, a Brother received 2,000 guilders; with this sum he had to get by. Usually, the Brother spent 5-6-900 guilders on top of the 2,000. And now, each station costs 5,000-6,000 guilders.... In my view you have two choices: either you decide on a fixed sum again or you ask an experienced Batak carpenter for an offer. I am convinced that we would then build for 1,000 guilders, if not less. We have enough Bataks who have the skill to do this. However, we have to oversee construction, in order for the house to be correctly built. Even in the worst case, there would only be minor construction mistakes, but we would never have to cope with costs higher than 1,000 guilders. When a Batak decides to take on piecework, he is capable of acquiring much better, and cheaper, wood than we can.[34]

This quote makes clear that some missionaries either did not know how to contain building costs, or that they had no interest in doing so. More important, the quote indicates an abundance of local craftspeople who were knowledgeable and skillful enough to procure suitable materials at reasonable costs, and to construct buildings of reasonable quality. To ensure that houses were built to appeal to European tastes, however, missionaries were required to install themselves as site managers.

Preparing the Site, Beginning to Build

Fries' circulars provide us with stories that illustrate the points made in the previous section. These stories support the interpretation that missionary stations served as trading zones where Europeans and Asians exchanged information and negotiated successfully—despite their contrasting identities and worldviews.

Two years before Fries and the local carpenters built the small hut in Siwalubanua, Fries had laid the foundation for his own missionary station in Sifaoroasi. The station was situated in the Niasan highlands, several days' walk from Gunungsitoli, home to the harbor, the region's main settlement, and the trading hub. In early 1905, after several negotiations with the local chief, Fries managed to acquire a suitable piece of land in Sifaoroasi. Without further ado, he set out to look for carpenters and assistants. Within a couple of days, Fries had managed to "hire 30–40 Niasan wood workers, headed by a skilled *Ober-'toeka'*"—an innovative conjunction of the German word for "master" or "superior" and the Batak word

[34] Brother Wagner's letter, dated 20 Sept. 1913, is to be found at VEM: Archival No. 2.838 ("Boetar-Pormonangan, Sumatra").

for "carpenter" or "craftsman." The workers' first task was to prepare "the necessary wood (planks, beams, etc.)," a task they carried out without supervision from Fries. In the interest of troubleshooting, Fries and the carpenters had embarked on "long negotiations," in which they decided on the various sizes of the wood elements needed, and their exact price.[35]

The master carpenter's name was Ama Mbolitae. Fries was very happy to have been able to engage Mbolitae, a "nice and reliable" person, who had grown up in Ombolata:

> He has already overseen the construction of several stations, and he will now oversee the felling of the trees in the jungle (as well as the dressing of the timber on a piecework basis). Later, he will direct the construction work—for a little less than one guilder per day. I am really looking forward to working together with him, when we in due course build the house in accordance with the layout.

Before construction could start, however, the designated plot had to be readied. By offering tobacco to anyone who was willing to help to remove shrubs and dig a well, Fries managed to get "everyone in the *kampong* [village]" to work for him.[36]

The next step was to mark the boundaries of the planned buildings: one residential dwelling and a number of outbuildings. Once the wood had been delivered and prepared, construction could begin. Less than three months after he acquired his plot, Fries documents the intense activity on the site:

> Yesterday, on Saturday, more than 50 people were working under my supervision. To the right, 10 prepared the ditch for the bathroom; to the left 25 were clearing the ground; at the back 4 *toeka* working on the horse stable; at front 10 leveling off the road.[37]

Six weeks later, between 50 and 100 people were involved in constructing the roof, and the *toeka* could begin to lay the floor.[38] As time went by,

[35] Quotes in this paragraph are taken from Fries, *Rundbriefe 2*, 1905, Circular No. 16, p. 13.

[36] Quotes in this paragraph are taken from Fries, *Rundbriefe 2*, 1905, Circular No. 17, p. 21.

[37] Fries, *Rundbriefe 2*, 1905, Circular No. 18, p. 25.

[38] Ibid., p. 31.

Fries found himself more and more involved, both as a supervisor and a manual worker: "If necessary, I also become a carpenter."[39] Although he relied on the master carpenter, Fries was willing to get his hands dirty. During the construction of a school building in Siwalubanua a few years later, one visitor found Fries "carefully clambering up the roof framing ... to fix the ridge."[40]

It had proven comparatively easy to attract workers to the remote village of Sifaoroasi. More problematic was ensuring the regular supply of food and materials. The village was poor and could not provide "15 to 20 carpenters with rice twice a day." Fries had no choice but to have "rice sent from G. Sitoli 70 km away at enormous costs."[41] Fries also ordered from Gunungsitoli "tar to paint the piles and sand from the sea to cement the bath basin," along with various "iron items for house construction."[42] Despite these problems, the first buildings were ready by June of 1905. A drawing published in one of Fries' circulars displays the building's signature Niasan elements, including stilts; a high, grass-thatched roof; and a generous porch.[43] (Fries was not alone in hewing to local building traditions. Hendrik Kruyt, a Dutch missionary to Sumatra, dreamt of a house that featured a huge, boat-like roof, rendered in the Batak style.[44])

When the time came, Fries was happy to leave the "small hut" where he had lived while his new home was under construction. In the fall of 1905, after he had married Elfriede, the young couple moved into the new house. In line with traditional roles for European women, Elfriede Fries' tasks were to keep the home clean and orderly, to cultivate flowers, plant "Sumatran seeds," and to try her hand at growing "European potatoes."[45]

[39] Fries, *Rundbriefe 2*, 1905, Circular No. 16, p. 13.

[40] Fries, *Rundbriefe 5*, 1908, Circular No. 4, p. 29.

[41] Fries, *Rundbriefe 2*, 1905, Circular No. 17, p. 24.

[42] Fries, *Rundbriefe 2*, 1905, Circular No. 18, p. 26; No. 17, p. 19.

[43] Fries, *Rundbriefe 3*, 1906, Circular No. 26, p. 38.

[44] Smith Kipp, Rita, *The Early Years of a Dutch Colonial Mission: The Karo Field*. Ann Arbor: University of Michigan Press, 1990, p. 67.

[45] Fries, *Rundbriefe 2*, 1905, Circular No. 20, p. 43.

Working Together

Fries and Nommensen's experience of actively building the mission station was shared by others in their cohort. Missionary Ewald Krumm "had to work intensely at the construction site from morning to evening."[46] And, in a report home, Wilhelm Steinsiek in 1890 describes the demanding building activities in the Lake Toba area of Sumatra: "… the missionary always has to be around. Not only do we supply all iron items, nails, locks, etc. We also have to act as carpenters." Unlike their compatriots in the Niasan jungle who could outsource the task of acquiring timber, missionaries in the Sumatran highlands had no choice but to purchase wood personally, from local traders.

> Each Monday, the day of the great Onan, (we'd say market day), people bring timber from the mountains to sell. Half of the day, and sometimes longer, you have to measure and buy planks and beams for construction. Of course, the Society pays everything, but it is still exhausting.[47]

When they first set out, European missionaries brought with them nails and locks, tools, and cash for wages and materials. When necessary, they later imported these items. For example, to facilitate building schools in his district, Dutch emissary Jan Wijngaarden ordered sets of carpenters' tools from Europe.

Constructing schools was a central element in missionary societies' proselytizing strategy. The majority of teachers who ran the missionary schools were men from the local population, converts to Christianity whom the missionary societies then hired. Before teaching could commence, however, the men were required to organize building materials and make sure the schoolhouse was properly constructed. According to Wijngaarden, most of the teachers easily adopted European building tools. In many cases, local chiefs did their part to help by compelling villagers to gather building materials from the forest and assisting teachers during the construction process. Wijngaarden reports that even the chief of the Karo Batak village of Tanjong Beringin lent a hand when the local school was under construction. After all, no missionary society could erect a building or open a school without the consent of the local chief.[48]

[46] *BRMG*, 1899, p. 152.
[47] *BRMG*, 1890, p. 50.
[48] The information in this paragraph is based on Smith Kipp, *Years*, pp. 110–111.

The missionaries' standard narrative about the local population reflected the contemporaneous, morally repugnant view of non-European peoples: missionaries typically characterized locals as filthy, lazy, and thievish. In service to carrying out their so-called civilizing mission, clergymen saw it as their task to convert heathens to Christianity as well as to teach them European values and customs.[49] Given that missionaries and their wives depended on the indigenous population to serve as maids, cooks, stable boys, craftsmen, and other laborers in their daily life, missionaries also had considerable self-interest at stake when it came to inculcating locals with European ideas and practices. For example, Nommensen's sense of satisfaction is evident in his account of local boys in Sumatra learning to work with European tools:

> The heathens need concrete role models. To illustrate: I started to labor heavily at my workbench, and several boys sat there and watched me; I didn't ask anyone to help me, but it didn't take long until they asked me to hand over the axe and wood plane. Now I only need to show them how it works.[50]

Historical sources also include examples of local chiefs taking the initiative to learn European techniques—or sending others to acquire that knowledge. For example, Brother Hanstein reports home that he had recently "talked with the man [a chief's son] for a long time; he had several questions, especially concerning house construction. He was accompanied by a joiner who was supposed to make chairs for him."[51] Displaying the standard, demeaning manner of the era, Brother Krumm mentions another chief, whose "greediness knew no limits" after he spotted the missionary's toolbox.[52]

Despite the locals' interest in European tools, the missionaries continued to rely on the knowledge and expertise of indigenous carpenters. Even Nommensen, a somewhat skilled craftsman, acknowledged that he

[49] Concerning the role of the "civilizing mission" in the history of science and technology, see Moon, Suzanne, *Technology and Ethical Idealism: A History of Development in the Netherlands East Indies*. Leiden: CNWS Publications, 2007.

[50] *BRMG*, 1863, p. 135.

[51] *BRMG*, 1882, p. 209.

[52] *BRMG*, 1899, p. 152.

knew very little when it came to "joinery, blacksmithing, roof construction."[53]

It is important to contextualize the locals' openness to learning how to operate European tools and to incorporate European building materials, such as iron nails. This openness and even enthusiasm did not impel indigenous craftspeople to give up their centuries-old building techniques. For example, some missionaries imported handsaws to help expedite the work of building, but this particular tool held little appeal: "People here still use adzes [an axe-like tool] to make planks. Since they consider sawing a tiresome procedure, they prefer to cleave the trunks."[54]

Research from the second half of the twentieth century shows that local carpenters upheld their traditional methods, for the most part. Anthropologist Wolfgang Marschall spent nine months in Hilizolagötanö, a village in the southernmost part of Nias. During this field trip, Marschall observed various aspects of the local material culture, including carpentry. By the 1970s, the locals had warmed to the idea of using saws, but they continued to shun metal items when constructing a house: "no peg, no screw, and no nail." Beams were held together by joints, and other parts of the structure were tied together with ropes or twigs. The advantage was that a house could be taken apart within half a day—and rebuilt at another site within three days. Marschall found hammers, chisels, and knives among the carpenters' tools—as well as, indeed, saws.[55]

Information Exchange

Historical sources support the proposition that missionary stations developed into "trading zones." One the one hand, the floor plan of the missionary dwelling mirrored European ideas of how to organize family life. On the other hand, practically the rest of the home reflected the materials and the construction methods embedded in the cultures of Nias and Sumatra. And, while local carpenters learned how to use a wood plane, they would continue to cut and finish planks in the same way as their fathers had done. During the planning and construction phases, missionaries and craftspeople collaborated, creating buildings that were neither

[53] *BRMG*, 1863, p. 135.

[54] *BRMG*, 1888, p. 244.

[55] Marschall, Wolfgang, *Der Berg des Herrn der Erde: Alte Ordnung und Kulturkonflikte in einem indonesischen Dorf*. Munich: Deutscher Taschenbuch Verlag, 1976, p. 89.

European nor Asian. These hybrid designs were the outcome of communication and negotiation between "foreigners" and locals.

It would be inaccurate to conclude that indigenous craftspeople merely carried out the orders issued by the missionaries. Europeans and locals did far more than negotiate wages and discuss the size of beams; they also worked design solutions together. Such negotiations did not always follow expected patterns: in Pangaloan, Sumatra, for example, Missionary P. Bonn noted that his congregation convinced him to use wooden shingles instead of reed and grass for the roof of the renovated church.[56] Some missionaries may have initially envisioned their homes as projecting a German style; soon, however, they realized the advantages of covered porches and steeply pitched roofs. Both design solutions were adjusted optimally to the regional climate: these adaptations allowed air to circulate and protected both the structure and its inhabitants from the ravages of tropical rains.

I consider it important to emphasize that the missionaries to Sumatra and Nias depended on locals for their labor and their knowledge. In many cases, villagers devised technical solutions that European missionaries could not have imagined:

> In one of the wards a couple of buildings had to be built. Since the site is underneath the level of a river, a high dam had been constructed. The idea was to erect the building on an elevated place. First of all, such a place had to be readied, but it proved impossible to find suitable stones, soil, and sand in the vicinity. However, the people knew how to find a way. They put a hollow trunk in the dam and diverted the river to a rice field, which had been bought for 50 guilders. There, the water was held back until sand and soil had settled. Then the water was led off, and the procedure continued until the plot was high enough. Then, a teacher's house—and later the school—was built on the elevated site that had been designed in this manner.[57]

Five years later, in the same area on Western Sumatra, ordinary people "willingly hauled sand, lime, and stones" for the construction of a water conduit.[58] The advantage: residents of the missionary station—as well as

[56] *BRMG*, 1890, p. 39.

[57] *BRMG*, 1907, pp. 237–238. The technique of constructing the truss and the roof first and moving the whole building to its final site is still used today in areas of Indonesia; see Waterson, *House*, p. 80.

[58] *BRMG*, 1912, p. 169.

the rest of the village—now had access to fresh water. Both the school and the water pipeline served the locals and their chief; apparently, this was reason enough to easily organize for collective labor.

Interestingly, the people reputed to have carried construction material "willingly" were women. Batak and Niasan women's work was not limited to the household. Women labored in the fields, prepared food in front of their homes, and traded goods at markets. Tasks linked to conventionally understood vocations—like carpentry or blacksmithing—were considered male domains, however.

Elfriede Fries and other missionaries' wives seldom carried stones and sand. Their task was to manage a European-style household and, once a station had been established, to adopt the role of teacher of practical subjects such as needlework. Daily chores kept the housewife fully occupied, and she had to be prepared to carry out many tasks herself.

Given that missionary women's voices rarely made their way into the Rhenish Missionary Society reports, historians have turned to other sources for these insights. Diaries, letters, and autobiographical sketches articulate the difficulties these (mostly young) women experienced in their foreign environments. While expectations remained almost the same as at home in Europe, preconditions posed serious challenges. For example, when other missionaries visited a station, they expected food and drink—preferably *European* refreshments. So, servants from the local population had to be taught to cook German-style food. And given that water was not always available directly on the premises, someone was obliged to haul it to the house. All meals had to be meticulously planned, and there was a garden to tend. Whereas fruits, vegetables, and chickens could be bought locally, lamp oil, canned foods, and so-called colonial goods had to be procured from stores that were often many miles—and several-days' journey—away.

To contend with these circumstances, European women communicated and traded goods and information with the locals. One missionary's wife, who lived at a station in India, describes how she had to "negotiate for long stretches of time" with hawkers who offered "fabrics, buttons, ribbons, shells, pearls, gems, curiosities or oranges, hens, etc."[59] Insofar as their language abilities allowed them to, European women instructed

[59] Quoted from Konrad, Dagmar, *Missionsbräute: Pietistinnen des 19. Jahrhunderts in der Basler Mission*. Münster: Waxmann, 2001, p. 296.

maids and other employees on how to use European household items and how to deal with the manually driven (nonelectric) washing machine.

Missionary wives endeavored to learn various techniques from local women: how to dehull and pound rice, how to process yams, how to grind coconuts and treat chilis, and how to grow indigenous vegetables. Just as missionaries' houses acquired a hybrid character, so did their households. For example, when white potatoes were in short supply, the missionary wife might have had no choice but to serve the roasted ham with sweet potatoes instead of "European potatoes." Johanna von Erlen, stationed on the west coast of Nias, had learned how to be pragmatic: "here, everything does not have to fit together perfectly."[60]

Despite the challenges, missionaries' wives strove to prepare dishes that matched European menus as closely as possible. Johanna Diehl's diary from her life in German New Guinea (the northern part of today's Papua New Guinea) yields detailed information about menus and cuisine. In 1907, Diehl had joined her husband-to-be, who, five years earlier, had been dispatched by the Rhenish Missionary Society to Bogadjim (a settlement in the Astrolabe Bay). A typical meal included "bean soup, then canned meat with beans and at the end a sweet pudding." On their wedding day, the couple offered guests a cold buffet which was just as lavish as if they had married in Germany. After "chicken broth in cups," there was "roast goose, boiled ham, potato salad, cheese, bread, white wine, and cake, (apple pie, sheet cake, and raisin cake) and cookies, various fruits (cherries, apricots, cucumbers), along with cider and wine and beer." On a more typical day, the Diehl family enjoyed either boiled hen, ragout, or jellied meat, served with potato salad, for example. Common desserts were apple pie, and pancakes with lingonberries. A typical pantry would contain canned "Frankfurt sausages." Johanna Diehl owned Weck jars, which allowed her to can vegetables, fruits, and meat. She churned her own butter and ground her own mustard. She made her own liverwurst and meatballs. She baked currant bread and macaroons, and made her own marzipan.[61] As much as possible, Diehl maintained German culinary traditions.

Johanna Diehl benefitted from living close to a harbor, where the German New Guinea Company regularly delivered goods from Europe

[60] Quoted from Töpperwien, *"Gehülfin,"* pp. 54–55.

[61] The information and quotes in this paragraph are taken from Klein, Dieter, ed., Jehova se nami nami*: Die Tagebücher der Johanna Diehl, Missionarin in Deutsch-Neuguinea 1907–1913*. Wiesbaden: Harrassowitz Verlag, 2005, pp. 10, 13–14, 35, 46–47, 63, 65, 74, 96.

and other parts of the world. Still, the Diehl family had to make compromises: given the absence of grapes on New Guinea, Diehl made wine from pineapples, instead. After some experimentation, she discovered that it was possible to make flavorful soups and fritters out of "*Sambi*, a kind of potato."[62] Although Diehl had a stove in her kitchen, she allowed her maid to cook over an open fire, in front of the house.

A closer look at cooking and cleaning practices supports the notion of missionary stations as figurative trading zones that merged East and West. Elise Eisfelder, who had married a missionary dispatched by the Swiss Basler Missionary Society, provides glimpses of life at a station in Sumaddi in western India in 1895. Pupils at the mission boarding school followed a strict daily routine. At six in the morning, they were hustled out of bed, and the older students prepared "rice porridge," which was served at six-thirty. At noon, the same children had to grind "grain for the bread."[63] In accordance with a widespread practice in rural Indian, the pupils once a week applied cow dung to plaster the mud floor. This scenario exemplifies how European ideas of temporal discipline and hygienic practices were combined with local customs.

As long as they were convinced of a domestic technology's usefulness, European women who lived at missionary stations readily appropriated that "new" technology. This was the case for Deborah Hoch, a German émigré and missionary wife. In the 1890s, Hoch lived on the southwestern coast of India. Confronted with the problem of how to dry the family's laundry during the monsoon season, Hoch adopted—

> ... an appliance that nobody knows in Europe. In a tin cabinet a cord is stretched, on which the wet clothes are hung. Below the clothes, you make a fire, in one or two earthen bowls filled with coal. You close the cabinet door and after two or three hours all the clothes in the box are nice and dry. Such a box is particularly helpful when you have small children and need to wash many diapers every day.[64]

From Sumatra, we learn that missionaries put the legs of their larders in bowls on the floor. To prevent ants making their way to the provisions, the bowls were kept filled with water.[65] The willingness on part of the Europeans to adopt local solutions is apparent.

[62] Ibid., p. 55.

[63] Quoted from Konrad, *Missionsbräute*, p. 291.

[64] Quoted from ibid., p. 295.

[65] Winkler-Metzler, Luise Emilie, "Meine Erlebnisse und Erinnerungen an die Zeit in Sumatra von 1905 bis 1921," at VEM: Archival No. 1-040443b, p. 77.

Appropriating Local Foods

Just as they rapidly adopted particular local building materials and technologies, most European missionary families seem to have quickly learned to appreciate domestic vegetables, fruits, and culinary specialties. Author Luise Emilie Winkler-Metzler documented the home life—including the foodways—at a missionary station in an unpublished autobiography titled, "My Experiences and Memories of My Time in Sumatra from 1905 to 1921." The book describes how indigenous spices "gave the sauce a wonderful taste," as well as how "European vegetables" in the family garden grew alongside "'*gadong hau*' [manioc], a green, somewhat bitter leafy vegetable, the dick and white roots of which one could also fry." The property also boasted ***antadjau*** [a type of guava] trees, the fruits of which tasted "delightful."[66]

Winkler-Metzler's autobiography is a cornucopia of information about everyday life at and around missionary stations on Sumatra. The author was a thoughtful, reflective observer of daily practices. While we cannot rule out the occasional misremembered name or date, we can largely rely on the accuracy of her detailed descriptions of foodstuffs, food-preparation techniques, and individual dishes.

As with every archival document, however, Winkler-Metzler's manuscript requires a critical reading. Luise Emilie Metzler was born in the Dutch East Indies. At the age of five, she was sent "home," to attend school in Germany, while her parents remained on Sumatra. In 1905—merely one-and-a-half years after Eduard Fries arrived on Nias—at age twenty-two, Luise Emilie reluctantly returned to the missionary station in the Batak village of Pearadja; as an "obedient daughter," the young Luise Emilie had little choice but to join her parents.[67] Three years after her return, she married Johannes Winkler, a German medical doctor, also stationed in Pearadja. (Johannes Winkler would go on to publish extensively on Batak ways of healing.)

Winkler-Metzler sees missionary life through the lens of sentimental memory. More important, the author is relatively unaware of the cultural appropriation—and the cultural oppression—of the "civilizing mission" in which she was engaged. After all, Winkler-Metzler wrote the memoir from the safety of her home in the university town of Tübingen, Germany; she

[66] Ibid., pp. 20–21, 24, 37.

[67] Ibid., p. 2.

likely saw herself as recounting the story of her life in "exotic" Southeast Asia.

For Europeans, life in Pearadja held at least one advantage over life at Lake Toba in the highlands. The station lay fairly close to Sibolga—a larger settlement and harbor—which was well connected to the wider world. Once a week, mail arrived from other parts of the globe. The village boasted a well-stocked "department store," which belonged to a Dutch company, Toko Hennemann.[68] For those who could afford it, the shop made it possible for missionaries to access almost everything they could have dreamed of. In this way, Pearadja was directly connected to the wider world.

Local products also abounded on Sumatra, and daily life could have scarcely functioned without these domestic goods. The main source of foodstuffs for Europeans and Asians alike was indeed the weekly market. In this physical, commercial trading zone, people of many ethnicities and nationalities negotiated and transacted business. In her signature romantic tone, Winkler-Metzler describes the "shouts and screams and bargaining" at the market in Tarutung, where everyone—including Dutch and German, Chinese and Japanese, as well as Batak people—tried to make themselves understood:

> In the burning heat, rows of women were sitting on the ground, selling fruit, beans, or cucumbers that they had planted in their own gardens, in addition to chili peppers, sweet potatoes, *tiung* [eggplant], a Batak kind of vegetable. The men dealt with tobacco and hens.

In addition, sellers offered rice, coffee, bananas, coconuts, and dried fish, as well as freshly slaughtered pigs and buffaloes. In the years immediately preceding the First World War, the Tarutung market grew considerably, as more and more East Asians opened stores on the site. Their inventories included bakery products, canned goods, and even artisanal craftwork.[69]

In addition to markets, gardens and kitchens developed into sites where people gathered to communicate and exchange information. The Metzler family kitchen was a figurative trading zone where culinary traditions and practices merged in a way that is reminiscent of today's "fusion cuisine," further discussed in Chap. 6. Luise Emilie's mother planned meals and

[68] Ibid., p. 9.

[69] Quotes in this paragraph are taken from ibid., pp. 21, 88.

discussed preparation techniques with her local cook, much as Fries and Nommensen negotiated with local carpenters to find design solutions. In the family garden, "European" and Asian vegetables coexisted. Given that neither indigenous people nor resident Europeans cross-bred plants, it would be inaccurate to call the garden a "hybrid" one. The presence of potatoes and tomatoes—plants of American origin—could, however, qualify the Metzler family garden as a global site.

The Metzler family developed a taste for tropical fruits, as they are called today. Some of these were ready-to-eat; others required preparation. For example, the family ate eggplant and papaya (*botik*) with sugar only, but found it necessary to boil the guava fruits. Luise Emilie explains that the fruit of the *djambu* (myrtle) tree required cooking with "water, sugar, and a small amount of vinegar."

> "Zuurzak" [soursop] was one of the most wonderful fruits. It was ... the size of a baby's head and had a firm, prickly skin. In the middle, it had white pulp with brown pits the size of beans. We pressed the interior [flesh of the fruit] through a sieve and added a bit of sugar. It made a gorgeous cream with such a pleasant taste that I cannot describe it to you.

The preparation of arrowroot as a thickening agent was more "cumbersome work." "The white roots had to be thoroughly cleaned and pounded. The same instrument [a large wooden mortar (*losung*) and pestle] was used to mash the arrowroot as to pound rice." Next, the mashed arrowroot was washed several times and pressed through a sieve. Using a piece of cloth, the arrowroot mash was then wrung out. Finally, it was set to dry. Given that wheat was not cultivated on Sumatra, arrowroot was especially useful as a thickener "for puddings, baby pap, or soups."[70]

The fruits that Winkler-Metzler mentions were often served after the family and guests had indulged in "Reistafel." *Rijsttafel*, an elaborate meal with Dutch colonial roots, is comprised of many small dishes. This grouping of dishes, which has erroneously been identified as "Indian," was (and still is) served with rice and "exotic" spices. The resulting food combinations of, for example, rice, chicken, and curry sauce, symbolize, if not epitomize, the Dutch appropriation of Indonesian culinary elements.[71]

[70] Quotes in this paragraph are taken from ibid., pp. 25–26, 27–38.
[71] Quotes in this paragraph are taken from ibid., p. 20, 37.

Thereby, we return to a topic which has been hovering above the narrative in this chapter: the asymmetric power relationship between the local population on the one hand and the Dutch colonial administration and the missionary societies on the other hand. The fact that the trading zones which I have been discussing were based on mutual information exchange may not obscure the fact that construction sites and kitchens were based on strict divisions of power—as well as on strict divisions of labor.

CHAPTER 3

Communicating and Trading in West Africa: Talking Drums and Pack Animals

> The monkey proposed that they make drums, an idea that had come to him in a dream while he was speaking with the ancestors in the next world. The monkey explained his solution like this: "It is not really my idea; it is really the ancestors" [*sic*]. They instructed us to cut down some trees, to cut the trunks into smaller pieces, and to hollow out the insides. The ends should be open. One end should be covered by the skin of a cow. [...]
>
> The lion, king of the animals, ordered that the monkey's idea be done. First loggers and sculptors had to be designated to cut the trees and shape the drums. The buffalo, the gazelle, the hippopotamus, the chimpanzee, the civet and the wild boar all volunteered.
>
> Other animals were enlisted to bring meals to the sculptors. Kakou Ananze, the spider, was asked to bring them palm wine.[1]

This excerpt comes from an African fable titled "The Origin of the Talking Drums." According to the tale, "the source of this invention" lies in the animal world, where, "a long time ago, [...] the animals of the jungle formed their own society, and it was separate from the society of men." The animals lived in harmony with one another, although they occasionally needed to gather "to discuss problems of common interest."

[1] Cangaley, Manny, *The Voice of the Elders: An Evening of Storytelling with the Baoulé.* Cologne: Rüdiger Köppe, 2005, pp. 147–149.

M. Hård, *Microhistories of Technology*, Palgrave Studies in the History of Science and Technology,
https://doi.org/10.1007/978-3-031-22813-1_3

To solve complex problems, they needed to contact their ancestors. But how could they "inform all the animals, spread throughout the jungle, about a meeting to be held in their jungle village?" Given that other solutions had proven inadequate, the lion decided to accept the monkey's idea. The fact that the suggestion to make drums had come from the monkey's highly valued ancestors also played a role in the lion's decision.[2]

Making the drums proved to be easier said than done. The animals that volunteered for the task found it demanding:

> After several days of hard labor, they succeeded in cutting down a few trees. The trunks were cut, and then it took several more days to hollow them out. When this was done other animals were sent to the village of the men to buy leather skin. Finally, they shaped the drumsticks.

It was equally difficult to find a talented drummer, an animal that "could communicate with the living and the dead." Several animals proved to be incompetent; the hippo, for example, "was too slow and heavy." Only after consulting three oracles did the animals discover that the spider was "the designated drummer." So, the lion called on the spider—Kakou Ananze—to play the drums; his performance was an instant success. Standing on two of his legs, the spider used his other legs to play six drums simultaneously: "As by a miracle he played the drums as if he had been doing so for ages. [...] He produced a melodious and enchanting rhythm."

Unfortunately, the spider's success was short-lived. Whereas some animals were intrigued by Kakou Ananze's wonderful music and obvious talents, others envied the spider's sudden rise to fame, and objected to the spider's boastfulness and arrogance. "Over time the jungle was divided into the admirers and detractors of Ananze." In the end, the poor spider was actually murdered, and "a civil war broke out between the animal camps." Rather than unifying the animal kingdom, the drums had contributed to its demise. The talking drums were rendered silent.

Only after "a hunter named Atungblan" found the drums in the jungle and brought them to his village did the drums enter the world of humans. Addressing the "dumbfounded" villagers in the Baule language, the drums themselves explained what they could be used for:

[2] All quotes in this and the following three paragraphs are taken from "Story 9: The Origin of the Talking Drums," in: Cangaley, *Voice*, pp. 149–161.

> We can send our message over long distances to living things and even to the spirits in the next world. We can announce the nobility of kings and chiefs with our sounds. The combination of these sounds and a magical rhythm makes all things move to our music.

As the story goes, this information impelled the Baule people to make drums "talk." The only remaining task was to devise a name for them, and what could be more natural than "to name the drums after the person who had found them"—Atungblan? According to the fable, *atungblans* have been used for communication and entertainment ever since their discovery.[3] And, in fact, the Baule—like many other West African people—devised a drum-based communication system of their own. This entailed using different drums for different pitches, thus creating coded messages that could be received by people within earshot, though at a distance. Once received by drummers in surrounding areas, the message was then relayed to a widening circle of people.

Talking Drums as a "Technology"

For the Baule, talking drums were an important communication technology. In this chapter, I discuss this and other kinds of technology used by the Baule and other African peoples before their lands were formally colonized toward the end of the nineteenth century. My approach challenges the received notion that equates "technology" with machines and so-called advanced scientific knowledge.[4] Before colonization, African peoples did not use the electric telegraph, for example, yet they enabled themselves to communicate over long distances, via drums.

Similarly, the Baule people did not design railroad lines, yet they created extensive trade routes across the continent. They did not make use of mechanical devices (except firearms), yet they succeeded in gathering foodstuffs and producing textiles and other goods for a larger—and even a global—market.

For this chapter, I have focused on sources from West Africa, especially from the area eventually called Côte d'Ivoire by the French, and Ivory

[3] Gourlay, K.A., "Atumpan," in: *Oxford Music Online*. Oxford University Press: doi.org/10.1093/gmo/9781561592630.article.L2232253 [accessed 7 August 2020].

[4] Schatzberg, Eric, *Technology: Critical History of a Concept*. Chicago: University of Chicago Press, 2018.

Coast by the British.[5] I have also used sources from contemporary Ghana and Nigeria. The environments analyzed here are those in which indigenous peoples communicated, traveled, transported, traded, cultivated, extracted, and manufactured. To contribute to a more accurate, more inclusive concept of "technology," I go beyond the analysis of mechanical, electrical, and digital techniques, referring to the environments in which these activities took place as "technological landscapes."[6] Indeed, the forests, the savannas, and the coastal areas where Ivory Coast dwellers (Ivorians, as they are now called) lived were not only natural landscapes but also technological ones. On the one hand, these natural landscapes provided inhabitants with raw materials for their technologies: wood, metals, and caoutchouc, for example. On the other hand, inhabitants transformed these raw materials into tools, utensils, and weapons, with which they modified their ecosystems. It was these resources and these technologies that enabled the inhabitants to live off the land.

The Talking Drum as Tradition

The previously excerpted fable about the origin of talking drums was codified and translated into English by scholars with knowledge of Baule culture. A subgroup of the Akan people, the Baule (*Baoulé*) live primarily in the central part of Côte d'Ivoire. Given that the Baule have not always employed written language, memory and the oral tradition have been—and remain—important cultural elements. As in most cultures, storytelling is an important vehicle for conveying knowledge and mores. Stories can take many different forms, and they can be performed in various ways. Among the Baule, stories are typically told during leisure time, "when the

[5] David Drengk has kindly provided me with the printed and archival sources on which the rest of this chapter is based. The ideas I present here have been developed in close discussion with Mr. Drengk, whose doctoral dissertation has the preliminary title "People, Technology, and Nature in Everyday Life: Changes and Continuities in the Technological Landscape of the Evergreen Rainforest in Lower Ivory Coast, c. 1890–1930."

[6] As discussed at the end of this chapter, I borrow the concept of "technological landscape" from historian of technology Svante Lindqvist but redefine it to serve the line of argumentation indicated in this paragraph; see Lindqvist, Svante, "Changes in the Technological Landscape: The Temporal Dimension in the Growth and Decline of Large Technological Systems," in: idem., *Changes in the Technological Landscape: Essays in the History of Science and Technology*. Sagamore Beach, MA: Science History Publications, 2010: 3–24. The text was originally published in 1994.

community gathers after dinner."[7] Most parts of the talking-drum fable were recited, and some parts were intended to be sung. Ideally, the audience was to take part in the singing.[8] Toward the end of the tale, the moral of the story becomes explicit: "Everyone has special talents. One must be tolerant and forgive. You must never be mean, you must never hate, and most of all you must never kill."[9]

Historians of Africa are keenly aware of the importance of oral tradition in passing on evidence, information, and values. Indeed, many topics in African history cannot be understood thoroughly without consulting oral sources. Historians and other scholars differentiate between the information gleaned from an interview and the information transmitted via oral traditions.[10] Whereas interviewees are able to give us insights into events the interviewees themselves have experienced, oral traditions "have passed from mouth to mouth, for a period beyond the lifetime of the informants."[11] Those who tell the "moral fable" about the origins of the talking drum are not inventing the main story line for the first time.[12] The storyteller is at liberty to modify certain passages, and perform the story with a different level of drama, though the main message and structure of the tale remain the same.

The story of the talking drums, which belongs to the literary genre of the fable, is an integral element of the Baule people's oral tradition. The story's animal characters are personified, the animal kingdom an allegory for human society. As with similar tales, the fable emphasizes a moral—rather than historical facts. One of the facts the story does convey, however, is that "talking" drums have long played a central role in Baule society. Accordingly, the storyteller specifies that talking drums were introduced generations ago: "Our grandparents inherited them from our great-grandparents, and they in turn, received them from our distant ancestors."[13] As it were, "The Origin of the Talking Drums" is an age-old morality tale—and a story about the drum as a real-world technology.

[7] Cangaley, *Voice*, p. 14.

[8] Concerning the practice and form of oral storytelling in Baule culture, see Cangaley, *Voice*, Introduction.

[9] Cangaley, *Voice*, p. 159.

[10] One of the pioneering texts in this strand of scholarship is Vansina, Jan, *Oral Tradition as History*. Madison, WI: University of Wisconsin Press, 1985. See also several of the contribution to Philips, John Edward, ed., *Writing African History*. Rochester, NY: University of Rochester Press, 2005.

[11] Vansina, *Oral Tradition*, p. 13.

[12] Vansina, *Oral Tradition*, p. 82.

[13] Cangaley, *Voice*, p. 147.

The Drum as Artifact

Outside observers of African culture were often impressed by the communication power of the "talking" drum. Albert Bushnell Lloyd, a missionary to Uganda, marveled that, "from one village to another, a distance of over a hundred miles, a message could be sent [via drums] in less than two hours."[14] Similarly, the infamous explorer Henry Morton Stanley was struck by the precision of the talking drum during one of his colonizing missions to what would become the so-called Congo Free State. Recalling a visit in 1883 to the Wagenya people, island inhabitants in the Congo River, Stanley wrote that they "have not yet adopted electric signals, but possess, however, a system of communication quite as effective. Their huge drums, by being beaten at different parts, convey language as clear to the initiated as vocal speech."[15] Indeed, Stanley astutely related talking drums to the telephone or the telegraph.

In 1949, Baptist missionary John F. Carrington published a book-length analysis of talking drums in the northeastern part of today's Democratic Republic of the Congo.[16] Describing how the Lokele people harnessed drums for communication purposes, Carrington shows how skilled drummers simulated the pitch of the Lokele language to produce understandable signals. Despite the differences between various African languages, Carrington suggests that the languages spoken in most of the sub-Saharan regions of West Africa lent themselves to this purpose. Baule oral tradition provides evidence of Carrington's statement: *atungblans* could indeed be employed to transmit messages based on the Baule language—a subgroup of the so-called Kwa language group.

Carrington differentiates two kinds of drums. The first category includes the skin-topped drums described by the monkeys in the Baule fable. The second category includes "all-wooden drums, hollowed out through a longitudinal slit."[17] For either of these drum-types to be used as a means of communication, they must enable drummers to produce high and low pitches. This pitch-based communication system is analogous to

[14] Quoted from Carrington, J.F., *Talking Drums of Africa*. New York: Negro Universities Press, 1969 (orig. 1949), p. 9.

[15] Stanley, Henry Morton, *The Congo and the Founding of Its Free State: A Story of Work and Exploration*, Vol. 2. Cambridge: Cambridge University Press, 2011 (orig. 1885), pp. 158–159.

[16] Carrington, *Talking Drums*.

[17] Carrington, *Talking Drums*, p. 21.

Morse code, which employs long and short signals to convey messages. In the case of skin-topped drums, two drums of different size and character are employed, whereas all-wooden drums are designed to allow the drummer to produce different sounds by striking different sections of the drum. Lokele drummers employed wooden sticks—usually with rubber tips made from vines found in the forest.[18]

The fable of the talking drums tells us they were initially invented to help the lion summon his animal-subjects to important meetings. Lokele drummers used their instruments to convey messages of different kinds. Crucially, drums were used to issue a simple warning of approaching danger to neighbors; but drummers could also send messages with far more nuanced content. Drum signals could inform dispersed community members of the new moon, the death of a certain person, or an upcoming social event. Drums could also be employed to summon the party needed for a planned hunting expedition. Drummers even sent personal messages, such as professions of love. And, importantly, drums were used to contact ancestral and other spirits.[19]

Drums also provided musical accompaniment to key social gatherings, such as parties and sporting events. Again, the Baule fable captures allegorically how the drum functioned as entertainment:

> The sounds made all the animals sway. Even the lion, king of the jungle, could not help nodding his head to the magical rhythm of the drums. He smiled broadly. The monkey clapped his four hands in appreciation of Ananze drumming. The tail of the leopard twitched to the beat of the drums. The elephant's huge ears fanned back and forth in time with Ananze's drumming.
>
> As for the serpent, his body coiled and turned. The gnat danced sometime [*sic*] on the back of the buffalo, sometime [*sic*] on the back of the elephant. The hare's teeth chattered and his ears twitched to the beat of the drums.[20]

Large Technological Systems in Colonized Spaces

Talking drums served various purposes, not only communication, but also worship and entertainment. The electric telegraph, by contrast, was monofunctional. By the time Stanley traveled up the Congo River, telegraph

[18] Carrington, *Talking Drums*, p. 28.
[19] This paragraph is based on Carrington, *Talking Drums*, pp. 57–71.
[20] Cangaley, *Voice*, p. 153.

lines were developing into a globally spanning network.[21] After some initial glitches, the first intercontinental telegraph lines became operational in the mid-1860s, incorporating Europe, North America, British India, and North Africa. Two decades later, the British West African Telegraph Company and the Spanish National Submarine Telegraph Company established services from Europe to several outposts and towns along the west coast of Africa. Grand-Bassam, the old commercial hub of what would become Côte d'Ivoire, was one of these towns. Despite its successive colonization of West Africa from the mid-nineteenth century onward, the French government continued to depend primarily on the services of British telegraph companies until the early twentieth century.[22]

The expansion of telegraph lines exemplifies technology's vital role in the processes of colonization and globalization. Along with steamships and railroads, telegraph lines contributed to what nineteenth-century economists and public intellectuals called "the annihilation of space and time."[23] Daniel R. Headrick is one of the historians who has analyzed the role of communication and transportation technologies in the imperial project. Together with powerful weapons, pharmaceuticals, and steamships, telegraph systems made up *The Tools of Empire*, as Headrick frames it, as one of his book titles.[24]

At first glance, the "tools-of-empire" thesis appears intuitively convincing. After all, steamships and steam locomotives sped up traffic and made travel more predictable; machine guns enabled colonial forces to quash most forms of military resistance; and quinine helped Europeans to stay healthy and survive inhospitable environments. Indeed, European technologies appear to have made colonial powers' "control and coordination" efforts easier to realize.[25] On deeper consideration, however, colonial

[21] Müller, Simone M., *Wiring the World: The Social and Cultural Creation of Global Telegraph Networks*. New York: Columbia University Press, 2016.

[22] This paragraph is based on Headrick, Daniel R., *The Tentacles of Progress: Technology Transfer in the Age of Imperialism, 1850–1940*. New York: Oxford University Press, 1988, Ch. 4.

[23] The use of this parable in connection with railroad traveling in Europe was impressively analyzed by Schivelbusch, Wolfgang, *The Railway Journey: Trains and Travel in the 19th Century*. New York: Urizen Books, 1979 (orig. 1977).

[24] Headrick, Daniel R., *The Tools of Empire: Technology and European Imperialism in the Nineteenth Century*. New York: Oxford University Press, 1981.

[25] Headrick, *Tentacles*, p. 107; Wenzlhuemer, Roland, "The Telegraph and the Control of Material Movements: A Micro-study about the Detachment of Communication from Transport," *Technology and Culture* 58 (3), 2017; 625–649; here: p. 642.

administrators in far-away countries—the so-called "men on the spot"—learned how to tweak and elude incoming telegraph messages, using this technology to their own advantage. In addition, as this chapter will show, colonial officers were not the only people who managed to evade attempts at colonial control: members of the local population also managed to do so.

In part inspired by this "tools-of-empire" narrative, historians of technology in colonial settings have tended to investigate large technological systems and networks as means to increase the influence of imperial powers.[26] Steamers, railroads, and telegraph lines dominate accounts of the nineteenth century; electricity networks, roads, and water-provision systems all feature in twentieth-century narratives.[27] Several studies point to the technological and commercial continuities between the colonial and the postcolonial phases.

In reality, the success of the large, powerful technological systems and networks was often limited. For example, given their large size and deep draft, British steamers were unable to navigate far upstream the Indus River, and traders remained dependent on smaller vessels and local boatsmen. Many technological visions never materialized. British imperialist Cecil Rhodes's wild idea of building a railroad line from Cairo to Cape Town was one of those rogue visions; starting in South Africa, the railroad line never made it further than the Copper Belt, in today's Zambia. Further, in most African colonies, large technical infrastructures remained fragmentary: whereas nationwide railroad and telegraph networks developed in British India and the Union of South Africa, technological systems never reached universal coverage in sub-Saharan colonies. Railroads typically went from the coast to important commercial centers or extraction sites, sometimes including a couple of tributary tracks. For a long time, electric-power plants worked in "island mode," with few or no

[26] "Instruments of Empire" is the title of Chap. 4 in Misa, Thomas J., *Leonardo to the Internet: Technology and Culture from the Renaissance to the Present.* Baltimore, MD: Johns Hopkins University Press, 2004.

[27] Straeten, Jonas van der, *Capital Grids: A Global History of Electricity in East Africa.* New York: Palgrave Macmillan, forthcoming; Sousa, M. Luísa, "Colonial Centres and Peripheries: Low-cost Roads and Portuguese Engineers in the 1950s," in: Fari, Simone, and Massimo Moraglio, eds, *Peripheral Flows: A Historical Perspective on Mobilities between Cores and Fringes.* Newcastle upon Tyne: Cambridge Scholars Publishing, 2016: 169–188; Akala, Jethron Ayumba, *In the Technological Footprints of Urbanity: A Socio-political History of Water and Sanitation in Nairobi, 1899–2015.* Darmstadt: Technical University of Darmstadt, 2019. https://tuprints.ulb.tu-darmstadt.de/8550/ [accessed Aug. 19, 2022].

connections to other plants. Rather than developing into integrated systems, technical infrastructures remained linear or isolated, for the most part.

Despite the fragmentary state of technical infrastructure in colonial settings, historians—whether of technology, of economics, or of the environment—continue to investigate infrastructure design and development. Although historians have begun to realize the limited power that large technical infrastructure had in practice, scholars still tend to gravitate toward large-scale, modern technologies. One reason for this bias is the fact that large infrastructure projects produced an array of (comparatively) easily available archival sources. In essence, many historians spill more ink on the telegraph than on the talking drum. As such, for the rest of this chapter, I will compare these large-scale technologies to precolonial ones that West Africans have used for centuries—many of which have remained unaffected by colonial projects.

Transportation Networks in West Africa

In 1906, the governor of Côte d'Ivoire, François Joseph Clozel, characterized the French government's technological engagement as "penetration" of the country. This technological campaign began with telegraph lines. Whereas ten years earlier only 200 kilometers of telegraph lines had been laid out (primarily along the coastline), that figure had now increased tenfold. Clozel proudly announced that the Ivorian network had been connected via a northwestern line to two other French colonies in West Africa: today's Mali and Guinea.[28] Slowly but surely, the French government established its proprietary colonial communication network.

The government's next huge task would be to design a railroad line from the Ivorian coast into the interior of the colony. Under the direction of military engineer C.F. Maurice Houdaille, construction commenced in early 1904, and the first 79 kilometers of tracks were finished by the end of 1906. The line began in Abidjan, on the coast, and terminated in Ery-Macouguié, a village just south of Agboville. A second construction phase began one year later, and in 1912 the tracks reached the town of Bouaké, the center of Baule country, situated a little more than 300 kilometers from the coast.

[28] Clozel, F.-J., *Dix ans à la Côte d'Ivoire*. Paris: Augustin Challamel, 1906, p. 11.

In contrast to the extensive, globally integrated telegraph system, the pioneering railroad was nothing more than a single line that traversed the natural landscape. Despite Houdaille's ambitious plans to connect various railroads across the vast area renamed French West Africa, an elaborate network never materialized.[29] Houdaille died in 1916; only four decades later did his envisioned railroad line reach Ouagadougou, the capital of today's Burkina Faso. Even in the 1950s, after more than a half-century of colonial power, the *Afrique-occidentale française* could not boast a comprehensive railroad network.

Governor Clozel had remained modest about his achievements—wisely so, in retrospect. Although he lauded the "progress achieved" in the field of infrastructure development, Clozel confessed that while the French government could design "great works, railroads, harbors, roads, canals," it could never guarantee that its efforts would lead to "increased production" or the development of new "needs." Reflecting the racist, colonialist ideology of his time, Clozel expressed his hope that the French government's efforts would support "the domination and civilization of the savage people or the barbarians of French Africa."[30]

Clozel admitted that the natural landscape of West Africa proved a serious obstacle to the government's planned exploitation of the colony:

> It is impossible to use steamers on its rivers. Even steamers with very shallow draft can navigate only short distances from the mouths of the rivers (around sixty kilometers, even for those best suited to the job). Further upstream, the strong rapids always make it difficult for canoes to navigate; sometimes it is dangerous, and at times, even impossible. In addition, in three-quarters of the country, the beasts of burden are in poor condition, and can barely carry out useful work: pack mules, donkeys, horses, oxen. Finally, two-thirds of the Ivorian region are covered by tropical forests, a terrain that causes numerous accidents and makes it more or less impossible to rely on pack animals—even if we were able to acclimate them.[31]

Like Stanley in Congo, Clozel pinned his hopes on the rapid expansion of the railroad. In the meantime, however, Clozel and his cohort recognized the need to rely on caravans with human porters and pack animals.

[29] Houdaille, M., *Le chemin de fer et le port de la Côte d'Ivoire: Organisation d'une entreprise coloniale.* Paris and Nancy: Berger-Levrault, 1905.

[30] Clozel, *Dix ans*, pp. 6–7.

[31] Ibid., pp. 13–14.

Postponing their faith in the progressive power of modern, mechanical technologies, the French colonialists in West Africa thus continued to depend on—and to expand—centuries-old, regional and local transportation practices. To this end, the French government, by 1906, had cleared more than 3,200 kilometers of footpaths throughout the country. In Northern Nigeria, the British made a similar move, extending the use of pack mules in the first decade of the twentieth century.[32]

Indigenous West Africans had always transported goods either by foot, with the help of pack animals; or in canoes and small boats. Trade routes traversed the region, from the coast to inland areas, and even across the Sahara Desert. For example, the Muslim city of Kong, in the north of what would become Côte d'Ivoire, was connected by human and animal transportation networks with Ougadougou, with Kano (in today's Nigeria), and with Bamako and Timbuktu (in today's Mali). From Timbuktu, further routes led to Tunis, Tripoli, and other towns in North Africa.[33] According to a conservative estimate from a late-nineteenth-century source, "twelve thousand camel-loads of goods" crossed the Sahara from North to South and back each year.[34] Toward the southeast, the network included Salaga and Kumasi (in today's Ghana) and outposts on the so-called Cape Coast, where Africans and Europeans had been trading goods—and tragically, slaves—for centuries.[35]

Clearly, a vast transportation system existed in West Africa well before the French built their first railroad line. This preexisting system connected practically every village and town. Rather than a network based on steam power and machinery, this system included human, natural, and material elements. In total, this is what I refer to as the precolonial technological landscape of West Africa.

[32] Ogunremi, Gabriel Ogundeji, *Counting the Camels: The Economics of Transportation in Pre-industrial Nigeria*. New York: NOK Publishers, 1982, pp. 76, 111.

[33] Bauer, Kerstin, *Kleidung und Kleidungspraktiken im Norden der Côte d'Ivoire: Geschichte und Dynamiken des Wandels vom Ende des 19. Jahrhunderts bis zur Gegenwart*. Berlin: LIT, 2007, p. 60.

[34] Robinson, Charles Henry, *Hausaland, or Fifteen Hundred Miles through the Central Soudan*. London: Sampson Low, Marston and Co., 1896, p. 119.

[35] The literature on the transatlantic slave trade is enormous. One of the most influential works dealing with the early phase of this inhuman trade from an African perspective is Law, Robin, *The Slave Coast of West Africa, 1550–1750: The Impact of the Atlantic Slave Trade on an African Society*. Oxford: Clarendon, 1991.

The precolonial transportation system was primarily a trade network (though warlords and kings also used the network to exercise power, much as the French later instrumentalized the railroad).[36] The most important merchants in the area surrounding the northern Ivory Coast town of Kong were the Dyula (Jula). The Dyula were a Muslim people who spoke Manding, and had moved to Kong from the west, in the fifteenth century.[37] The Dyula carried out both local and long-distance trade across the savanna, from the perimeters of the rainforest in the south, to the borders of the Sahel region. In the north, the Dyula purchased salt, wool fabrics, and horses, in addition to iron tools, weapons, and household items. From the south, the Dyula acquired gold, textiles, herbs, and kola nuts, as well as European muskets, munition, and liquor, among other goods and commodities.[38] Enslaved people were treated as high-priced trade goods.

Compared to the railroad, the traditional transportation system was slow. The pace depended on both the topography of the landscape and the season. Travel time was generally shorter during the dry season than during the rainy season. Long-distance trade was often dangerous and unpredictable; heavy vegetation and wild rapids made certain routes impassable. To survive, traveling merchants were required to know how to acquire foodstuffs along the way—even in sparsely populated areas.[39] In peacetime, they traveled alone or in small groups; during periods of conflict, traveling merchants formed large caravans that could number several hundred. The caravan was either armed, or it sought protection from regional warlords. Many merchants had their own slaves, who carried loads up to fifty kilos on their heads.

Not all porters were enslaved people. Traders' children also carried heavy loads, as did women traders. Some traders hired professional porters, though, apparently, it was not a widespread practice. This situation changed with the growing needs of European explorers, merchants, and colonizers in the area.[40] By paying carriers a salary, albeit a low one, Europeans contributed to the creation of portage as a part-time occupa-

[36] Şaul, Mahir, "The War Houses of the Watara in West Africa," *The International Journal of African Historical Studies* 31 (3), 1998: 537–570.

[37] Ibid., p. 544.

[38] Bauer, *Kleidung*, p. 59.

[39] Binger, (Louis-Gustave), *Du Niger au Golfe de Guinée par le pays de Kong et le Mossi*, Vol. 1, Paris: Librairie Hachette, 1892, p. 131.

[40] Ogunremi, *Counting*, pp. 72–76.

tion. Like the railroad, professional porters became "tools of empire," so to speak. Horace Tremlett, a woman who in 1914 led a party of roughly fifty porters in the British Colony and Protectorate of Nigeria, reports that the "Government provides carriers for any traveller who gives notice of requiring them."[41]

According to numerous travel accounts, this transportation system, relying as it did on human portage, presented Europeans with certain problems and inconveniences. The diary of Charles H. Robinson, a British subject who in 1894 traveled extensively in the region populated by the Hausa people in today's Niger and Nigeria, is filled with accounts of the problems he claims to have experienced when hiring and managing large groups of porters. Well aware that White men and women depended on their labor, porters were often able to successfully negotiate their salaries and to set the pace of Robinson's party.[42]

The precolonial transport network cannot be considered a high-tech system, by any means. The system did involve material techniques, however. The paths themselves were mainly the result of "the padding of countless bare feet" rather than the deployment of heavy machinery.[43] Tree trunks were turned into bridges, and canoes were employed to cross deep rivers. Along frequently traveled routes, ferrymen offered their services. To mitigate the pain that came from carrying extremely heavy loads on their heads, porters placed their cargo on specially designed pads. Tremlett, who traveled in the same area as Robinson, describes this protective piece of gear:

> [Carriers'] most cherished possession seems to be the greasy little pad that they attach to the loads to protect their heads. It is composed of folded cotton, sewn up like a pincushion, and is tied on by various odds and ends of string, grass and strips of leather, its proper adjustment every morning being a matter of much careful deliberation.[44]

In addition, porters carried along a stick with a fork-shaped end, which they used to facilitate loading and reloading. Given the long transportation times, goods had to be well wrapped. Robinson relates how merchants made sure that kola nuts did not dry or deteriorate: "During the

[41] Tremlett, Horace, *With the Tin Gods*. London: John Lane the Bodley Head, 1915, p. 73.
[42] Robinson, *Hausaland*.
[43] Tremlett, *Tin Gods*, p. 75.
[44] Ibid., p. 73.

march the nuts are packed in baskets and covered with fresh green leaves. Every four or five days they ought to be re-packed, in order that the leaves may be renewed, and that the nuts which are touched with mildew may be removed."[45]

Although pack animals played a role in the West African transportation system, animals were not omnipresent, except in the North. Donkeys were sometimes used in the savanna and the forest during the dry season, though the tsetse fly always posed the serious threat of infecting all pack animals with trypanosomiasis ("sleeping sickness"). In areas where it was necessary to cross wide and deep rivers, for example, the use of pack animals was problematic.

Regardless of whether or not we consider pack and draft animals to be technological artifacts—a question hotly debated by historians of technology and environmental historians—donkeys and camels belonged to the technological landscape of the region.[46] Tremlett observes that "some days there was a good deal of traffic on the roads. Strings of donkeys padded patiently along with their burdens of sugar and cotton stuffs."[47] Horses also belonged to this landscape, although they were used to signal prestige and for warfare, rather than for transporting goods.[48]

Camels, too, were important for trade in the semiarid Sahel region and further north. John Leyden, an early-nineteenth-century European adventurer, noted that the camel was "an instrument formed by nature for effecting a communication across these immense wastes"—the Sahara Desert.[49] It would be a long time before any manufactured vehicle, like the truck, would be able to compete with this natural "vehicle," the camel. In the absence of roads that crossed the desert, experienced guides used the stars for navigation and followed established waypoints. As in the case

[45] Robinson, *Hausaland*, p. 116.

[46] Edmund Russell posed this question in 2001, thereby setting off an intense email-forum debate among environmental historians and historians of technology: "'Are Animals Technology?' Archive of Messages Posted on Envirotech," https://www.envirotechhistory.org/wp-content/uploads/2007/05/animaltech.pdf [accessed Aug. 19, 2022]. Cf. McShane, Clay, and Joel A. Tarr, *The Horse in the City: Living Machines in the Nineteenth Century*. Baltimore, MD: Johns Hopkins University Press, 2007.

[47] Tremlett, *Tin Gods*, p. 90.

[48] Law, Robin, "Horses, Firearms, and Political Power in Pre-colonial West Africa," *Past and Present* 72 (1), 1976: 112–132.

[49] Leyden, John, *Historical Account of Discoveries and Travels in Africa*, Vol. II. Edinburgh: George Ramsay, 1817, p. 493.

of human portage, loading technologies were essential. One historian describes this procedure:

> A pack-saddle was used, consisting of two sheaves of grass or straw, two semi-circular pieces of matting made of plaited dum [doum] palm fronds, a skin filled with grain or stuffed with dry camel dung and wooden arch terminating in flat boards. Then, the loads, which must be carefully balanced, are slung over the pack-saddle; two loops on each load are hitched to the other two with two short sticks. Girths or breastbands were used if the loads were bulky or needed special steadying.[50]

If strong enough and well packed, a camel could carry up to 250 kilos.

Trade and Commerce

Military officer Louis-Gustave Binger encountered many caravans and merchants during his journey of more than two years (1887–1889) through the southeastern parts of so-called French Sudan and the British-named Gold Coast. Binger observed that it was seldom possible for Dyula traders and other merchants to cover a distance of more than twenty kilometers per day. Given that beasts of burden could not pass everywhere, and their cargo often had to be reloaded, the use of pack animals did not increase speed. Pack animals were also expensive. Seldom more than twenty or thirty centimeters wide, paths were too narrow to accommodate wagons. Binger thus concluded that human portage was both the cheapest and fastest transportation method for the region. Dyula people who lived in Kong typically made one or two journeys annually. A typical trip might be from Kong to Djenné (Mali), via Bobo-Dioulasso (Burkina Faso), and back—nearly 700 kilometers each way.[51]

In his 500-page travel account, Binger, who would later be promoted to the role of Côte d'Ivoire's first governor, details the existing, precolonial trading routes.[52] The extent of the precolonial transportation arteries may surprise today's readers. One route, nearly 800 kilometers long, led from the trade center of Kankan in today's Guinea to Kangaré and N'Tentou in today's Mali—and all the way to Macina, close to Djenné, in the Sahel region. Binger notes the main commodities along this "large

[50] Ogunremi, *Counting*, p. 103.
[51] Binger, *Du Niger*, pp. 88, 127, 298, 318.
[52] Ibid., Ch. III.

commercial route": "from the north horses and salt, and from the south came kola nuts and slaves."[53]

Binger also documents the many villages and towns connected by the network. Designated locations along the most frequented routes had turned into small settlements where locals offered food, lodging, and storage facilities to traders and travelers. Whenever possible, Dyula merchants stayed with Muslim hosts, so-called *jatigiw*, the "owners of the shadow"—those who provide the trader with shelter.[54] Shared religious values created trust, especially important when merchants employed their hosts as translators, commissioners, or middlemen.[55]

The strenuous journey Binger undertook was not an innocent excursion. Financed by the French government, the trip was motivated primarily by commercial and military interests. Binger's travel report is filled with detailed information about the price of many kinds of goods, in various towns and at central markets. He was convinced that, for example, French textiles would be able to compete successfully with allegedly "inferior" British and German products on the West African market. Binger claims to be able to sell French calico fabrics in Kong at a 500 percent profit margin over the cost of the goods in Paris. Binger also asserts that if the French were to establish trading posts in Bondoukou and Groumania—half-way between Kong and the ports in Grand-Bassam and Assinie on the southern coast—Dyula merchants would "immediately create a safe route with lodges along the way," abandoning the British-controlled route to Salaga and Kumasi in the Gold Coast.[56] Marcel Monnier, a French journalist and writer, who in 1892 accompanied Binger on another trek through the area, claimed that such a route would enable traders to travel from the Ivorian coast to Kong within one month.[57]

The bustling town of Kong, with its commercial center, astounded Binger. He estimated the population to be 15,000 inhabitants, and he was taken aback by Kong's huge market square:

> In the center of town, you find the marketplace, about 500 meters long and 200 meters wide. Since its five or six trees do not provide enough shade,

[53] Ibid., p. 130.

[54] Ibid., p. 321; Bauer, *Kleidung*, p. 69.

[55] Bauer, *Kleidung*, pp. 69–72.

[56] Binger, *Du Niger*, p. 322.

[57] Monnier, Marcel, *France noire (Côte D'Ivoire et Soudan)*, Paris: Librairie Plon, 1894, p. 210.

> several merchants have built fairly comfortable straw-hut stalls, in which they spend the day during large markets.[58]

In the late afternoons, more than a thousand people gathered at the marketplace to buy and sell goods from near and far. Goods from the Sahel region and beyond included fresh meat and fruit, honey and sweets, kola nuts and chilis, millet and corn, yams, shea butter, cotton and wood, as well as salt. European products included hats, beads, tableware, and guns. According to Binger, some of the salt for sale at the Kong market (in today's Côte d'Ivoire) had been mined in Taoudenni (Mali), more than 1,500 kilometers away.[59] Kong was also famous for its textiles, especially its red-and-black-striped products, known as *el-harrotáfe*. Cotton was grown in fields outside the town.[60] Weavers sat in alleys, working at their mobile looms, making products such as colorful ribbons, which were highly valued. Binger counted 150 dyeing pits, each measuring approximately one meter in diameter and two meters deep—and emitting a "very strong odor."[61] Merchants often manufactured their own fabrics, carrying their looms with them on trade journeys. Kong was also an intellectual center and spiritual destination for Muslims, attracting pilgrims from great distances.[62]

Manufacturing Textiles, Searching for Gold

The Kong market reflected the richness of West African material culture. Social anthropologist Kerstin Bauer emphasizes Dyula merchants' active role in fostering cultural and technological exchange.[63] During their extensive travels, the Dyula transported and traded goods across cultural and geographic borders; they also contributed to the spread of design knowledge and skills, especially in the area of textile production. For example, Binger claims that some of the fabrics manufactured by the Dyula in Kong were attempts to imitate fabrics from Ségou, 600 kilometers to the North, in today's Mali.[64] But the textile-manufacturing learn-

[58] Binger, *Du Niger*, p. 297.
[59] Ibid., pp. 315–318; cf. also Monnier, *France noire*, p. 210.
[60] Bauer, *Kleidung*, pp. 103–104.
[61] Binger, *Du Niger*, p. 297.
[62] Bauer, *Kleidung*, p. 104.
[63] This paragraph is based on ibid., pp. 163–176.
[64] Binger, *Du Niger*, p. 317.

ing process also worked in the opposite direction: according to Côte d'Ivoire's governor, Clozel, the Abron people in the town of Bondoukou "had learned how to make cotton fabrics from the Dyula."[65] Similarly, evidence shows that Ashanti (Asante) weavers appropriated various textile techniques and patterns from the Dyula. Bauer shows that the mutual exchange of design and techniques among various West African peoples continued—and probably intensified—in the course of the twentieth century. For example, Dyula and Baule weavers and dyers borrowed textile patterns from each other during the colonial period.

Textile manufacture and trade held a prominent place in the technological landscape of northern Côte d'Ivoire. Cotton was grown in the area; elaborately woven textiles were in high demand. Traditionally, weavers were men; their most commonly used technology was the foot-treadle loom. Compared to most European machines at the time, West African looms were narrow—between two and thirty centimeters wide. Placed at the far end of the loom, a heavy stone was used to keep the warp threads taut; the warp could be several meters long. When manufacturing blankets, as well as robes, loincloths, and other garments, the weaver cut the narrow ribbons to suitable lengths and sewed them together. The most common colors were indigo, red, blue, and white. Various plants were used to create indigo dye, while kola nuts could be used to create burnt-orange fabrics. Many products were checkered or striped; some had elaborate embroidery.[66]

Blacksmithing and ironmaking also featured in the technological landscape of the area. As the makers of tools, weapons, and jewelry, blacksmiths enjoyed high social status in their communities. In Oumalokho, a village near the border between today's Mali and Côte d'Ivoire, Binger records five active blast furnaces. Impressed by the ovens' "practical" design, the attentive workmen, and their appropriate tools, Binger describes how iron lumps weighing between forty and fifty kilos were batch-produced. To capture the drama of the production process, Binger included in his printed report a black-and-white drawing of the fiery furnaces and the hardworking laborers.[67]

Binger was one of many colonialists who documented the natural resources and commercial activities in West Africa—with the ultimate goal

[65] Clozel, *Dix ans*, p. 54.

[66] Bauer, *Kleidung*, pp. 130–135, 145–159.

[67] Binger, *Du Niger*, p. 260.

of exploiting them to their own advantage. Administrators and military men, adventurers and companies, shared an interest in the treasures of the region. Even scientists belonged to this colonial vanguard: biologists, geologists, and cartographers, among them. For example, Frenchman Gaston Joseph, while carrying out cartographic work in Côte d'Ivoire, was motivated to comment on the local workers' accomplished gold-extraction techniques. Like Karl Helbig's positive observations about Borneo (see Chap. 1), Joseph's descriptions of Côte d'Ivoire praised workers' skills:

> We have nothing to teach the indigenous people in the gold districts of Côte d'Ivoire. Although they do not know anything about geological and petrographic phenomena, they have carried out their search in a very good way. Considering the truly primitive tools they have at their disposal, their work is remarkable—especially when it comes to extraction. [...] The indigenous people have exploited all kinds of gold deposits: mines, valleys, riverbeds.

Joseph added that engineers employed by European mining companies had already discovered the skills of "these excellent workers."[68]

Joseph also recorded the gold-mining activities of various Baule subgroups. In the Kokumbo area, not far from the Bandama River, these Baule workers dug mines into the schistose (layered, coarse-grained) soil, rich in quartz, which can contain gold particles. Joseph identified three types of mines: deep shafts or wells; long trenches; and wide, relatively shallow cavities. Measuring approximately one meter in width, the narrow design of the mineshafts allowed workers to maneuver up and down without using ladders, simply by "leaning against the wall" and using their hands and feet to climb and descend. In areas rich in deposits, miners created an intricate system of "large rooms and galleries" underground. To his astonishment, Joseph learned that it was possible to find mines almost eighty meters deep.[69]

To excavate the quartz, workers used "a shovel with a 40 cm long, curved handle, a pickaxe with a straight handle of the same length, a crude pick—consisting of a stick with a sharp iron cone at the one end." During the extraction process, when they encountered prohibitively hard rocks,

[68] Joseph, Gaston, "Exploitation indigène de l'or en Côte-d'Ivoire," *Bulletins et mémoires de la Société d'anthropologie de Paris*, 6th series, 4 (3–4) 1913: 372–375; both quotes are taken from p. 375.

[69] Ibid., pp. 372–373.

workers first heated the schist by means of a bonfire. Then they poured water into the shaft, forcing the rock to burst apart. Workers in the mine-shaft collected the quartz in baskets made from raffia leaves; the quartz was then hauled to the surface by means of a rope, also made of material from the *Raphia* palm. After removing excess stone and soil, workers carried the quartz to their villages, where women crushed the quartz on large stone slabs. Then, they pulverized the stone and began the painstaking search for tiny gold particles.[70]

In addition to excavating for gold in schist, West Africans also searched for gold in alluvial rocks, especially in river sediments and fluvial sand. To extract "dust, granules, or nuggets" from these sediments, the local workers used simple, wooden pans, fifty to sixty centimeters in diameter. Baule women were among those most active in panning for gold. If lucky, a woman was able to extract one gram of gold during the course of one day.[71] A French military engineer, G. Crosson, claimed that the Attie people, who lived in the lagoon district in the south of Côte d'Ivoire, had found in alluvial sediments gold nuggets weighing "hundreds of grams"; this is almost certainly an exaggeration.[72]

Even the French observers who found the Ivorians' gold-extraction tools somewhat "crude," felt compelled to marvel at the delicate instruments local merchants used to weigh and assess the quality of the gold. Camille Dreyfus, who spent half a year in Attie country in the late 1890s, describes the so-called "gold book" (*diah*) used for examining and weighing gold.[73] The *diah* was not a conventional book, but a small "package of waxed cloth," in book form. Carefully wrapped inside was a scale, which Nzema (Apollonian) traders used to weigh precious metals and coins: "The pans of this balance-scale are attached to the beam by very thin threads. To weigh the gold, you use your thumb to raise the scales by means of another thread, making sure to keep all other fingers separated." In addition to this delicate balance, the *diah* also contained several "weirdly shaped weights," a small bowl, and a "copper spatula." A hen's feather

[70] This paragraph is based on ibid., p. 374.

[71] Ibid., p. 375; Joseph has taken these numbers from a French "administrator," a certain Mr. Cheruy.

[72] "Rapport sur les Attiés concernant la construction du chemin de fer par M. le Capitaine du Génie G. Crosson," 23 June 1899; in: Archives nationales d'outre-mer (ANOM), Aix-en-Provence, France: 1 T.P Box 17, File: 1. The quote is taken from p. 31 of Crosson's report.

[73] Dreyfus, Camille, *Six mois dans l'Attié (un Transvaal français)*. Paris: L. Henry May, 1900, pp. 41–46.

and a magnet were employed to remove impurities from gold dust before weighing it. A black stone was used to check the hardness of gold nuggets.[74]

Dreyfus records how sellers and buyers of gold deployed their personal gold books to control weights and degrees of purity. Transactions were usually slow, as negotiations involved lengthy bargaining and discussion. Completing a deal required sellers and buyers to share the same weighing system and adhere to common social rules. While local transactions were embedded in a set of cultural values and norms, the "weirdly shaped weights" reflected the global dimensions of the gold trade. The basis for the locally employed weights was the troy weight system—the very same system that was used in Britain. In Kong, Salaga, and London alike, one troy ounce equaled 31.1 grams, and the various weights the Nzema traders used were fractions of a troy ounce. For example, Dreyfus noted that the gold book typically contained half-ounce weights, and others weighing one twentieth of an ounce. Given that West Africans had traded gold with Europeans for centuries, the use of the troy system in the Gulf of Guinea preceded British colonialism.

Cultivating the Land

Binger observed how the indigenous people of West Africa used the landscape's natural resources: they hunted antelope and other animals; they gathered and processed the fruits, nuts, roots, leaves, and other parts of plants and trees; they grew an array of foodstuffs. To ensure their food supply year-round, villagers built myriad granaries and other storage huts.

In the region just north of Kong, yam was the most important cultivated plant. Sweet potatoes were also widely grown, as were a white form of millet and a red type of sorghum.[75] People also collected the pods of the African locust bean (*Parkia biglobosa*) as well as the fruits of the shea tree (*Vitellaria*). The seeds of these trees were processed to season sauces and to make shea butter. Reporting on the cultivation of different corn varieties, Binger claimed that they were all of "inferior quality." And, although the balsa-like wood of the baobab tree was unsuitable for heating purposes and for carpentry, it offered many other possibilities. Binger noticed that:

[74] Ibid., pp. 41–42.
[75] Binger, *Du Niger*, pp. 246–249.

> [...] people use its ashes as a fixative in indigo dyes and as potash for soap-making. The bark is used to make twine, ropes, nets, hammocks, etc. The leaves are employed as a condiment in all kinds of sauces eaten together with *tô* [a grain paste].
>
> In certain regions, the fruit shells are employed as bottles; [...] The white flour that you find in the non-ripe fruits is used in the preparation of certain indigenous dishes and in certain drinks, mixed with millet flour; the pits are cooked, dried, and pounded to make *kondoro*, a sauce used as a preservative.[76]

Several years after Binger recorded this information, military engineer Crosson made similar observations in a report on the Attie people. French engineers' interest in the Attie stemmed from the fact that the first railroad was planned to run through Attie country. Crosson's unpublished report documents the various forms of providing food and manufacturing clothes and tools in the area. The Attie cultivated cotton, corn, yams, manioc, sugar cane, peanuts, peppers, pumpkins, and three kinds of plantains. In the forest, they harvested caoutchouc (natural rubber) from both trees and vines. From palm trees they harvested fruit and sap, processing these raw materials into wine and oil. Most palm oil was consumed locally, but villagers who lived on the shores of the lagoons had set up "indigenous oil-manufacturing sites that work year-round" and sold "large quantities" of palm oil to the French factories and trading posts on the coast.[77]

French botanists, often commissioned by the colonial forces, documented the elaborate ways in which West African forest dwellers utilized trees and other plants. For example, indigenous people exploited different species of palm trees, and they had developed an array of techniques for using almost all parts of the tree—not just the fruit and wood. This phenomenon, among others, interested the botanists, and some published their observations on "applied botany and colonial agriculture" in journals.

Despite the botanists' elaborate descriptions, colonial companies focused primarily on extracting oil from the African oil palm (*Elaeis*) and harvesting as much wood as possible from the evergreen rainforest. In the nineteenth century, one type of palm oil—made from the fruit pulp—was used in Europe and North America to make soap and candles; it was also employed in the tin-manufacturing industry and as a lubricant. In the twentieth century, another type of oil—made from the palm kernels—was

[76] Ibid., p. 244.
[77] ANOM: 1 T.P Box 17, File: 1, p. 31.

increasingly used for cooking purposes.[78] Wood from the jungle was exported and also employed in colonial settings, where public works departments and private enterprises used, for example, the palmyra palm (*Borassus*) for various infrastructure projects: telegraph poles as well as bridge pilings and pillars.[79]

Auguste Chevalier was one of the "colonial scientists" who studied indigenous West Africans' use of palm trees.[80] In one article on the African palmyra palm (*Borassus aethiopum*), as well as in another set of observations about the raffia palm (*Raphia*), Chevalier delves into the different ways in which local populations used these trees.[81] The palmyra was plentiful in Baule country and in other areas of West Africa; Chevalier admiringly estimates there to be thirty to forty million palmyra trees in French West Africa. He cites a colleague who claims to have observed a palmyra forest, in the region known today as Mali, measuring nearly 100 kilometers long and three-to-four kilometers wide.[82]

Palm wine, also known as toddy, was the most widely circulated product that non-Muslim peoples in Western Africa obtained from the palm tree. The wine was made of sap from the buds rather than from the bark (as with tapped maple syrup in North America). According to a later source, the process involved three differently shaped knives and was quite labor-intensive.[83] During the high season, the "crude earthenware jars" that gathered the sap had to be emptied twice a day.[84] The sap from some trees is sweet, whereas the sap from other trees is less sweet and is usually allowed to ferment, thus increasing the alcohol content. Chevalier claims that the best palm wine was made from the African palmyra palm (*Borassus*

[78] Lynn, Martin, *Commerce and Economic Change in West Africa: The Palm Oil Trade in the Nineteenth Century*. Cambridge: Cambridge University Press, 1997.

[79] Chevalier, Auguste, "Le Borassus æthiopum de l'Afrique Occidentale et son utilization," *Revue de botanique appliquée et d'agriculture coloniale* 10 (108), 1930: 649–655; here: 652.

[80] Bonneuil, Christophe, "August Chevalier, savant colonial: Entre science et Empire, entre botanique et agronomie," in: Petitjean, Patrick, ed., *Les sciences colonials: Figures et institutions*, Vol. II. Paris: Orstom, 1996: 15–35.

[81] Chevalier, Auguste, "Nouvelles recherches sur les Palmiers du genre Raphia: (Suite et fin)," *Revue de botanique appliquée et d'agriculture coloniale* 12 (127), 1932: 198–213.

[82] Chevalier, "Borassus," p. 650, 652.

[83] Portères, Roland, "Le palmier ronier (*Borassus aethiopum* Mart.) dans la Province du Baoule (Côte d'Ivoire)", *Journal d'agriculture tropicale et de botanique appliquée* 11 (12), 1964: 499–514; here: pp. 507–508.

[84] Tremlett, *Tin Gods*, p. 85.

aethiopum), and he estimates that "a typical adult tree yields five to six liters of palm wine, for two months per year."[85]

When not used for wine production, the sap was sometimes processed into sugar. When properly cultivated, a single hectare of palmyra palm trees could yield 1,500 kilos of sugar annually. Referring to this high yield, Chevalier predicts that palmyra sugar will soon become a serious competitor to cane sugar—especially in those parts of the tropics where it is unprofitable to grow sugarcane.[86]

The hard wood of the male palmyra tree was used widely in construction: for slats, roofing laths and rafters, as well as posts, stilts, and stakes. The far softer wood and often hollow trunk of the female palmyra tree was used to make gutters, fences, and beehives. The leaves of the tree were deployed both as a roof covering, and as a key raw material for handicraft work—including for the manufacture of bags, baskets, mats, and fans. Petioles—the part of the tree's stalk that joins the leaf blade to the stem—were processed to make brooms and rope. Indigenous workers left virtually no part of the palm trees unused; even the delicate leaf-veins were made into string and fishnets.[87]

Several parts of the palmyra palm are edible; Chevalier notes that forest dwellers happily consumed the tree's fruit. In the early months of the year, people ate the seeds (embedded in the fruit), which contain a sweet jelly that tastes a bit like coconut milk. In June through September, when the fruit had ripened, it fell to the ground and could be gathered easily. (Chevalier assessed the fleshy pulp of the seeds to be relatively fibrous and tasting mildly of turpentine, though he still found the flavor appealing.) Even unripe seeds found their way into the African diet; they were simply toasted or grilled. The trees' young, fresh leaves and buds were also considered to be edible, as were the roots of the palmyra palm. If boiled in water, the roots were said to have anti-asthmatic properties. No wonder that the versatility of the *Borassus* fascinated Chevalier, the colonial scientist.[88]

The raffia palm is also native to Côte d'Ivoire. For the most part, forest dwellers used this type of palm in the same ways in which they used the palmyra tree. Chevalier documents twelve discrete uses for the raffia palm,

[85] Chevalier, "Borassus," p. 652.
[86] Ibid., p. 655.
[87] Ibid., pp. 652–653.
[88] Ibid., p. 653.

including: the sap for palm wine, the leaves for roofing material, and the wood for rafters. Interestingly, Chevalier notes that parts of the raffia tree's young leaves were mixed with cotton to make fabric. Along the coast, especially, the production of piassavas from the raffia tree was a lucrative enterprise. The petioles from the raffia palm were processed into red-grey fibers of roughly fifty centimeters in length. These fibers were in demand locally and regionally. To manufacture the fibers successfully, however, workers were required to treat the petioles with great care, harvesting the dry leaves directly from the stem and soaking them in lagoons (or other water sources) for a couple of days.[89] This technique, called retting, which is well known in many parts of the world, facilitates the process of separating the plant stem from the leaf's useful fibers.

Chevalier wrote these two articles in the early 1930s, when the colonial system of domination was already firmly established. As a state-employed botanist, Chevalier's goal was to increase and rationalize the exploitation of natural resources in the colonies; to this end, he consistently tried to appropriate as much botanic and agricultural knowledge as possible from the local population. According to a biographer, Chevalier early on acquired a "colonial habitus," and he eagerly supported French military and commercial interests in Africa. For example, Chevalier "played a decisive role in the debates about the exploitation of caoutchouc and the craze for cotton cultivation in [French] Sudan." And, "his botanical research on various kinds of wood in Côte d'Ivoire, and later in Gabon, created new opportunities for the exploitation of the forests." From around 1900 to the outbreak of the First World War in 1914, Chevalier traveled 50,000 kilometers throughout Africa, collecting 70,000 botanic specimens. Although he realized that the French could learn a great deal about the use and cultivation of indigenous plants from the local inhabitants, he never doubted that these peoples would in the end profit from the civilizing mission of the colonizers.[90] Chevalier never questioned the colonial project.

The Technological Landscape, Revisited

We have seen how West African peoples used the natural resources in their surroundings to make a living. They extracted metals; they gathered foodstuffs; they hunted game; they grew crops. At the same time, they

[89] Chevalier, "Nouvelles recherches," pp. 199–200.

[90] Quotations in this paragraph are taken from Bonneuil, "August Chevalier," pp. 17–19.

developed and applied their own technologies to process food, manufacture fabrics, make and transport goods, and communicate across distances. As I have shown in the case of the Attie people, professional hunters, smiths, weavers, and healers embodied the necessary knowledge and skill for these activities.[91]

I suggest that the totality of these resources and their application made up a "technological landscape." Importantly, this technological landscape existed before the advent of colonialism, and it—in part—continued to exist after the French officially created Côte d'Ivoire.

The way in which I employ the concept of "technological landscape" differs slightly from its original meaning; my thoughts on the concept have also been influenced by Francisca Bray's use of the term.[92] When historian of technology Svante Lindqvist introduced this term in the 1990s, he wanted to challenge the discipline in two ways. First, he criticized the obsession by historians of technology with novelty. Most works in this field of scholarship deal with new inventions, new developments, new gadgets. Second, Lindqvist argued that this bias has led us to overlook the many technologies that were established long ago, but still exert considerable influence in our daily lives. Since they have become ubiquitous in our society, these technologies recede into the background: electricity networks and water-supply systems, for example, become invisible elements of our modern "technological landscape." Historians of technology tend to focus on Thomas Alva Edison and other so-called system builders; historians tend to forget the later phases in the life cycle of such systems—exactly the phases in which they have established themselves and exert their greatest influence.[93]

I have presented a cross section of the varied "technological landscape" of West Africa, as it existed well before the arrival of European colonizers. Neither governor François Joseph Clozel nor engineer C.F. Maurice Houdaille created the technological landscape of Côte d'Ivoire; the landscape was there well before these men set foot on Ivorian soil. With the advent of the railroad from 1904 onward, the technological landscape partly changed, but it existed as such long before the French arrived—for example, in the form of an extensive network of footpaths and caravan

[91] ANOM: 1 T.P Box 17, File: 1, pp. 31–32.

[92] Bray, Francesca, "Flows and Matrices, Landscapes and Cultures," *ICON: Journal of the International Committee for the History of Technology* 22, 2016: 8–19.

[93] Lindqvist, "Changes."

routes. The creation of technological landscapes is not the privilege of colonial powers. The telegraph lines and railroad tracks Clozel and Houdaille helped to design never *formed* a technological landscape. At most, these telegraph lines and railroad tracks were superimposed on an existing technological world that was complete unto itself.

CHAPTER 4

Withstanding Globalization in Northern India: Farmers Make Sugar for Local Consumption

> It is well-known to those who take an interest in the sugar industry, that a very large proportion of the sugarcane grown in these provinces is crushed by the small country-made mills, which are marvellously cheap. These are all vertical mills, which type has been discarded by the present scientific world as some of the expressed juice falling down by gravitation is absorbed by the lowermost cane passing from the mill and lost. They are designed to be rotated by a pair of bullocks [steers], such as are found locally and give an extraction of about 50 per cent. From a broad economic point of view, the low extraction given by these mills, is a national calamity, but from the cultivator's point of view, it is "Hobson's choice." He would like very much to get more juice from his cane, but he could not afford to buy or hire a better mill and stronger bullocks, which would be necessary to get higher extraction; besides, the khandsari, to whom he sells the juice, objects to his taking too much out of the cane, because with low extraction, there is less gum and other objectionable substances, which he cannot deal with in his primitive way [...].[1]

This is an excerpt from a 1916 report in which Ram Richh Pal Sanghi expresses his sympathy with the poor farmer in the United Provinces (now Uttar Pradesh) of Northern India. Even if the individual sugarcane

[1] Hulme, William, and R. P. Sanghi, *A Note on the Improvement of the Indigenous Methods of Gur and Sugar Making in the United Provinces: A Report on the Government Experimental Sugar Factory, Nawabganj, Bareilly District.* Allahabad: Government Press, United Provinces, 1916, p. 1.

M. Hård, *Microhistories of Technology*, Palgrave Studies in the History of Science and Technology,
https://doi.org/10.1007/978-3-031-22813-1_4

cultivator had wished to increase productivity and efficiency, he did not have the means to do so. At most, he could join forces with other peasants and lease basic equipment on a cooperative basis. Given that most cultivators were heavily in debt to local moneylenders, however, even a cooperative association would seldom be able to lease, let alone to buy, "more efficient" machinery.[2] In practice, the poor peasant had no viable option but to continue to use the methods and techniques his parents and grandparents had used. Other technologies were simply too expensive—and often required knowledge and skill which smallholder farmers did not possess.

Sanghi was astonished by how well the "indigenous methods" accommodated local conditions. Although he had been hired by the United Provinces' government as a "Sugar Chemist" to develop new, more advanced technologies, Sanghi had to admit that such solutions were far beyond the reach of low-income cultivators. If the peasant were miraculously able to afford an iron mill with heavier rollers, he would need to buy "more powerful and consequently costly" steers as well. Why invest in expensive, stronger pack animals when the peasant's existing steers were well-adapted to a variety of tasks: not only to drive the simple, vertical mills that crushed the sugarcane, but also to plow the land and to transport goods? Theoretically, the optimal solution might have been an engine-driven iron mill, but this would have been far too expensive and required both expert knowledge and access to spare parts: "The oil-engine is a complicated machine; it has many parts of delicate construction and is never efficiently managed unless thoroughly understood."[3]

Visits to the villages surrounding the Government Experimental Sugar Factory in Nawabganj, where Sanghi was employed, helped him to appreciate the "primitive way" in which the cane juice was refined. Common practice among professional *khandsaris* (sugar manufacturers) and local peasants alike was to use the so-called *bel* extraction system to manufacture muscovado (*rab*), a dark-brown sugar with a high proportion of molasses:

> The "bel" [sugar-refining setup] contains usually five pans of the cheapest possible kind, which are built up on a furnace made of mud, housed in a mud-house with a roof of grass. One cannot conceive anything that would be less costly. After the season is over the pans are taken away and the mud-house is left standing to be often washed away during the rainy season.[4]

[2] Ibid.
[3] The quotes in this paragraph come from ibid., pp. 1–2.
[4] Ibid., p. 1.

The manufacturing method for muscovado was characterized by simplicity and flexibility. After the sugar had been boiled, the concentrated juice was "placed in large earthenware vessels about two feet high and one and a half feet in diameter."[5] Some of the molasses was then let off through a small hole at the bottom of the vessel.

When there was a demand for crystallized sugar, the *khandsari* could choose to further refine the muscovado by employing "a river grass called *sewar*" to draw off the remaining molasses. Finally, the sugar was "taken to the *pata*, a level floor surrounded by a low wall. The sugar is here dried in the sun and crushed by coolies walking on it and twisting their feet as they walk." Another local technique for making crystallized sugar involved even greater manpower: muscovado was placed in gunny bags (probably made of jute), and six or seven bags were "placed one upon another." A worker positioned himself on the top of this pile of bags and held onto a pole for balance. One meter up in the air, then, the "cooly ... sways backwards and forwards, and by this means most of the molasses is separated from the sugar."[6]

Assuming the role of rational scientist, Sanghi felt compelled to complain about the extreme inefficiency of these local production methods: if peasants and sugar manufacturers were to employ "modern processes," they would produce more than three times more sugar from the given amount of sugarcane. However, Sanghi also understood that modern methods of refining sugar were unaffordable:

> Any improvement of a permanent character would probably cost a hundred times the amount or more than what is spent on the temporary mud-house and arrangement of pans. The cultivators have no money to spend, and the *khandsaris* who are rich do not desire any change.

[5] Hulme, William, and R. P. Sanghi, *The Improvement of the Indigenous Methods of* Gur *and Sugar Making in the United Provinces*, Bulletin No. 82 of the Agricultural Research Institute, Pusa. Kolkata: Superintendent Government Printing, 1918, pp. 4–5. Both *A Note* and *The Improvement* were coauthored by Sanghi and William Hulme. There is strong evidence that Sanghi was the main author, and that Hulme, referred to as "senior author" in *The Improvement*, was granted coauthorship on the strength of his position as Director of the Experimental Sugar Factory in Nawabganj. Judging by their names, most directors of government research stations were British or of British decent, whereas assistants were more likely to be indigenous Indians.

[6] The quotes in this paragraph are taken from Hulme and Sanghi, *The Improvement*, p. 5.

Compared to industrial technologies, the established methods were "wonderfully cheap," and the vast majority of the rural population lacked the means to try out new techniques.[7] In fact, the "primitive" methods Sanghi described were appropriate to the economic, technical, environmental, and social conditions of the Indian countryside. Specifically, the boiling pans were easily accessible, as they were usually made by the village blacksmith; the *bel* hut was constructed by local laborers. The steer was similarly practical: the peasant used the same animal for sugar-grinding as for plowing and transportation. The varieties of sugarcane used in Northern India were also suited to the region: while plant breeders and researchers in the late nineteenth century had begun to develop varieties that, at least in theory, contained more juice, the varieties chosen had proven comparatively resistant to frost, animal attacks, and disease.

Indeed, the traditional system of production exhibited several characteristics usually associated with "appropriate technology."[8] The technologies of cane cultivation and sugar manufacturing required minimal investment; they were based on knowledge and skills accessible in the immediate community; their repair and maintenance could be carried out by means of locally available material; and they were well-suited to the physical environment and social milieu.

Sanghi realized it would be almost impossible to bring about an "improvement of the indigenous methods of Gur [raw sugar rich in molasses] and Sugar making in the United Provinces." Initially, improving the sugar-making process "seemed to be quite an easy thing to do," though the scientists soon modified their research goals. Instead of adopting high-tech solutions, they tried "to make the machinery and process as simple as possible, so that the superskilled management and highly technical supervision, so essential in large modern factories, would not be necessary." In pursuing this strategy, Sanghi and Director William Hulme hoped to persuade at least a few "small capitalists" to found sugar-making factories. There is reason to doubt the appropriateness of their strategy, however. Sanghi and Hulme advised prospective investors to acquire an implausibly elaborate and costly array of equipment: cane-crushing machines with eleven rollers; a steam boiler; crystallizers; clarifiers "with copper steam coils"; evaporators; and a centrifuge.[9]

[7] The quotes in this paragraph are taken from ibid., pp. 3, 5–6.

[8] There is a vast literature on the Appropriate Technology movement; see, e.g., Dunn, P.D., *Appropriate Technology: Technology with a Human Face.* New York: Schocken, 1978.

[9] The quotes in this paragraph are taken from Hulme and Sanghi, *Note*, p. 1, and idem., *The Improvement*, pp. 1, 9–12.

A Familiar History of Sugar

To understand the uniqueness of the Northern Indian case, it is helpful to revisit the standard history of sugar. For more than two centuries, sugarcane has been one of the world's most important cash crops. Like coffee shrubs, cocoa trees, and tea bushes, sugarcane is grown throughout large areas of the Global South—Asia, Africa, Latin America, and the Caribbean. As the term "cash crop" suggests, these plants have given peasants, landowners, and investors an opportunity to sell their commodities (coffee or cocoa beans, tea leaves, and sugar) on the open market. The trend toward increasingly globalized markets distinguishes cash crops and cash commodities from products grown for reasons of pure subsistence, barter, or sale at small, local markets. Cash crops deliver certain advantages to cultivators as well as presenting various drawbacks: great dependency on rapidly fluctuating global markets and globally acting food corporations, and the tendency to engage in practices that are detrimental to the environment.

This chapter tells another story. As I will describe, the Indian case shows that sugar can be associated with phenomena other than colonialism and slavery, capitalism and globalization. The history of sugarcane *cultivation* does not have to be a story of plantation economies, slavery, and indenture.[10] The history of sugar *production* and refining is not necessarily a story of the application of scientific knowledge, large amounts of capital, and expensive machinery. And the history of sugar *consumption* cannot be reduced to the global spread of white sugar only. Despite the expansion of global capitalism and industrial forms of production, local markets and craft-based manufacturing methods remained in use. Sugar may well have "changed the world," as the authors of a popular book on the subject suggest, but sugar did not affect fundamentally all corners of the world.[11] In Northern India, traditionally grown sugarcane varieties proved well-adjusted to local conditions, and a large-scale plantation economy did not develop during the time of British colonization. Production remained small-scale, and most customers continued to prefer muscovado and brown sugar to highly refined, white sugar. Bringing in the central concept from the previous chapter, we could say that the sugar-production

[10] Cf. Bosma, Ulbe, "Local Peasants and Global Commodities: Sugar Frontiers in India, Indonesia and the Philippines," *Österreichische Zeitschrift für Geschichtswissenschaften* 30 (3), 2019: 42–62.

[11] Aronson, Marc, and Marina Budhos, *Sugar Changed the World: A Story of Magic, Spice, Slavery, Freedom, and Science*. Boston, MA: Houghton Mifflin Harcourt, 2010.

methods which Sanghi describes were well integrated in the technological, physical, and social landscapes of Northern India. From the point of view of the Indian peasant, these methods represented an *appropriate* form of technology.

Standard histories of cash crops entail narratives of global connections and the exercise of brutal power; the history of sugarcane is no exception. Since the early days of globally acting trade companies, the sugar business has encompassed large parts of the world, and the cultivation of sugarcane has been accompanied by the subjugation of labor and nature. Historians have documented extensively the cruel deployment of African slaves on the sugar plantations of Brazil and the Caribbean.[12] From the mid-seventeenth to the mid-nineteenth century, sugar played an important role in the notorious trade routes between Africa, the Americas, and Europe. Abolition may have improved marginally the plight of laborers, but working conditions remained dreadful.

The large-scale, monoculture-focused sugar-plantation economy also spread, for example, to Mauritius, an island close to Madagascar, and Natal in Southern Africa. When slaves were no longer available, plantation owners began to acquire indentured workers in large numbers—not least from India.[13]

Andrew F. Smith is one of many authors who have documented the relevance of sugar in the global economy.[14] In his book *Sugar: A Global History*, Smith explains that sugarcane began its journey across the globe more than two thousand years ago. Based on available evidence, which is inconclusive, some archeologists believe that sugarcane from today's New Guinea spread to Indonesia, then on to India and the Mediterranean region. After the Europeans began to conquer the Americas, they brought sugarcane to the newly discovered continent via the Canary Islands and Madeira. After sugarcane became established as a plantation crop, refined

[12] See, e.g., Moitt, Bernard, ed., *Sugar, Slavery, and Society: Perspectives on the Caribbean, Indian, the Mascarenes, and the United States*. Gainesville, FL: University Press of Florida, 2004.

[13] Kumar, Ashutosh, *Coolies of the Empire: Indentured Indians in the Sugar Colonies, 1830–1920*. Cambridge: Cambridge University Press, 2017.

[14] Smith, Andrew F., *Sugar: A Global History*. London: Reaktion Books, 2015; Abbott, Elizabeth, *Sugar: A Bittersweet History*. London: Duckworth Overlook, 2009, provides a similar overview of the topic, whereas Schwartz, Stuart B., ed., *Tropical Babylons: Sugar and the Making of the Atlantic World, 1450–1680*. Chapel Hill, NC: University of North Carolina Press, 2004, focuses on the Early Modern Period.

sugar became a staple of colonial economies. This typifies the historiography of sugar with its story line of sugarcane, as a plant, circulating globally, and refined sugar becoming a global commodity. Indeed, throughout the centuries, sugarcane has proved to be a highly mobile plant that lends itself to cultivation in several tropical and subtropical regions. This mobility mirrors other cash crops, such as tobacco, coffee, rice, and tea. In recent years, the global circulation of certain plants—along with the knowledge and skill that accompany their cultivation and use—has been studied intensely by historians of science, technology, and agriculture. The volume *Rice: Global Networks and New Histories*, edited by Francesca Bray, is a case in point.[15]

The global mobility of crops is necessarily accompanied by the circulation of certain techniques and expertise. When the first Spanish settlers in the Caribbean decided to start producing sugar, they brought with them—from the Canary Islands—seeds and seedlings, as well as cane-crushing mills and skilled personnel.[16] Despite this preparation, the colonizers faced formidable challenges: they had to clear land, adapt to local conditions, secure a labor force, and transport the refined product to distant customers. In Europe, the voracious demand for brown and white sugar, molasses, and rum, easily justified the economic risks implicit in sugarcane cultivation and transatlantic transportation. Anthropologist Sidney W. Mintz, in his book *Sweetness and Power*, describes how the use of sugar came to accompany the consumption of tea, coffee, and cocoa.[17] In nineteenth-century Britain, the practice of ending a meal with a sweet "pudding" circulated even among the working class.

Estimates suggest that over the course of the nineteenth century, world sugar production increased by a factor of twenty-five. To satisfy consumer demand for sweeteners in the Global North, European scientists and plant breeders experimented with different varieties of beetroot. Their goal was to isolate a plant with a high enough sucrose content to merit commercial

[15] Bray, Francesca, ed., *Rice: Global Networks and New Histories*. New York: Cambridge University Press, 2015. Recently, Francesca Bray has expanded this network perspective to include several globally circulating crops in the project "Moving Crops and the Scales of History," supported by the Max Planck Institute for the History of Science in Berlin: https://www.mpiwg-berlin.mpg.de/research/projects/moving-crops-and-scale-history [accessed July 27, 2022].

[16] Mintz, Sidney W., *Sweetness and Power: The Place of Sugar in Modern History*. New York: Penguin, 1985, pp. 33–34.

[17] Ibid., Ch. 3.

cultivation. In the mid-nineteenth century, the sugar beet, as it was called, became a serious competitor to sugarcane. Unlike sugarcane, the sugar beet did not require a subtropical climate. That meant sugar beets could be grown in Europe and North America, close to the main markets of the world.

To put it briefly, the history of sugar, as it is usually told, is one of global expansion and circulation. It also entails a narrative about standardization. Although plant breeders used hybridization to develop hundreds of varieties of sugarcane, the commercial end-products were relatively uniform.[18] Sugar products made from Indonesian sugarcane did not differ substantially from that made from Cuban sugarcane. In principle, white sugar made from sugarcane bears the same characteristics as sugar made from sugar beets.[19]

Sugar manufacturing for a global market has fostered mechanization and ever-larger investments.[20] From the middle of the nineteenth century onward, the kind of machinery described by Hulme and Sanghi could be found in many corners of the world. In 1850, a contemporary observer referred to what he called the "Cuban Industrial Revolution"—a process that transformed Cuba into one of the world's largest manufacturers of both sugarcane and sugar products.[21] Over time, cane-crushing mills tended to become larger and more efficient; increasingly, steam-engines were employed to drive centrifuges and other machines; and the so-called vacuum pan—a kind of pressure cooker—replaced the "primitive" method of boiling the sugarcane juice in open pans.[22] Sugar manufacturing shifted from individual plantations to centrally located factories; the construction of railroad lines as well as other initiatives were undertaken to speed up the transportation of cane and refined products.[23]

[18] Ibid., p. 238.

[19] Fernández-Prieto, Leida, "Mapping the Global and Local Archipelago of Scientific Tropical Sugar: Agriculture, Knowledge, and Practice, 1790–1880," in: Manning, Patrick, and Daniel Rood, eds, *Global Scientific Practice in an Age of Revolutions, 1750–1850*. Pittsburgh, PA: University of Pittsburgh Press, 2016: 181–198.

[20] This is the main narrative in, e.g., Galloway, J.H., *The Sugar Cane Industry: An Historical Geography from Its Origins to 1914*. Cambridge: Cambridge University Press, 1989.

[21] Quote by José Luis Casaseca, taken from Fernández-Prieto, "Archipelago," p. 189.

[22] A detailed account of various mechanical and chemical methods of sugar refining is provided by Deerr, Noël, in *The History of Sugar*, 2 Vols. London: Chapman and Hall, 1949–1950, Ch. 33.

[23] We encounter this standard story line in, e.g., Smith, *Sugar*, pp. 38–40. The establishment of large central mills in Australia is described by Griggs, Peter D., *Global Industry, Local Innovation: The History of Cane Sugar Production in Australia, 1820–1995*. Berne, CH: Peter Lang, 2011.

An Alternative History of Sugar

Northern India did not follow the example of Cuba; in this chapter, I show how the history of sugarcane cultivation and sugar production took quite a different path in this part of the world. It was in this subtropical part of the country that the government set up the experimental station at which Hulme and Sanghi worked. As Sanghi himself indicates, peasants in Northern India were poor and had few possibilities for changing their set ways of refining sugar. I argue that the established, preindustrial, "indigenous methods of Gur and Sugar making" had proven *appropriate* to the local community—from an economic, social, and cultural point of view.

While India has for centuries been one of the world's largest producers and consumers of sugar, its sugar economy remains a predominantly domestic affair. As B.S. Baviskar, an expert on the history of sugar production in India, framed it in the early twenty-first century, "the Indian sugar economy... has remained immune to the impact of the world capitalist forces."[24] When it comes to both the production and the consumption of sugar, India in fact ranks first in the world.[25] Still, India exports less than fifteen percent of its total production. Despite its high production and consumption figures, India plays only a small role in the global trade of this cash crop. This is the case today, just as it was throughout most of the nineteenth century. The history of sugar in India is exactly that: a local, regional, and national history of an allegedly global product. To better focus the discussion, I will limit my narrative to the last one hundred years of British colonial rule in India.

To understand the Indian sugar trade, it is helpful to start with the consumer side. Most of the sugar consumed in British India was not refined, crystallized sugar; it was mostly members of the urban upper classes who exhibited "modern tendencies" by consuming white sugar with tea or coffee—or enjoying refined sugar in the form of confections—on a regular basis.[26] Indeed, the majority of Indian consumers, especially those in the countryside, preferred *gur*—also known as *gul*, *gud*, or *jag-*

[24] Baviskar, B.S., "Indian Sugar Then and Now: Power to the Peasants," in: Moitt, *Sugar*, pp. 22–36, here: 22.

[25] Cf. statistics provided by the London-based International Sugar Organization: https://www.isosugar.org/sugarsector/sugar; retrieved Aug. 20, 2022.

[26] This quote is taken from Datta, R.L., and Tinkari Basu, "Improvements on the Country Process of Making Sugar from Gur," *Department of Industries, Government of Bengal, Bulletin No. 42*. Calcutta: Bengal Secretariat Book Depot, 1930, p. 3. Cf. Ray, Utsa, *Culinary Culture in Colonial India: A Cosmopolitan Platter and the Middle-Class*. Cambridge: Cambridge University Press, 2015.

gery—a raw sugar with a relatively low sucrose content that is usually light yellow to brownish, depending on its molasses content. (*Gur* can also be found in other countries in Southern and Central Asia.) At markets, *gur* was sold in the form of semi-soft "cakes or balls or lumps," and it was a common ingredient in all kinds of Indian dishes—not only in desserts and sweets.[27] Consider the important differences between the two kinds of sugar and their fates. The market for *gur* and muscovado was primarily a domestic, even local one. In contrast, the market for white, refined sugar had direct connections with international trade. Accordingly, it was primarily producers and consumers of white sugar who felt the impact of price fluctuations on the global market.[28]

The manufacture of *gur* was as low-tech and affordable as the production of muscovado. Sanghi was impressed by the simplicity of the manufacturing process as well as by the sensory skills required. Laborers boiled the sugarcane juice in nothing but "a hole in the ground," and the fuel consisted of dried sugarcane leaves and stalks. "The attendant, when the mass in the pan becomes very thick, takes out a little, dips it into cold water, and rolls it into a small ball, and by this means he can tell when it is ready." Given the lack of access to thermometers and other instruments, the experienced *gur* boiler (*pakwa*) was required to employ his eyes, nose, and fingers to decide, for example, when the workers should "vigorously stir" the boiling liquid and when they should remove the boiling pan from the furnace. Once the *gur* had begun to cool down, the dough-like substance had to be worked over with a wooden mallet to reach the desired texture.[29]

Hulme and Sanghi were not the only scientists and engineers among their contemporaries who expressed surprise at the persistent use of *gur* and muscovado in Indian society. In 1930, Chemists R.L. Datta and Tinkari Basu described the cultural embeddedness of raw sugar:

> Gur itself forms an important item in the dietary of the inhabitants of India and by far the greater part of the total output of this commodity is consumed in the crude state, a comparatively small portion only being available for conversion into sugar. Under existing conditions good quality edible gur

[27] Hulme and Sanghi, *Improvement*, p. 4.

[28] Amin, Shahid, *Sugar Cane and Sugar in Gorakhpur: An Inquiry into Peasant Production for Capitalist Enterprise in Colonial India*. Delhi: Oxford University Press, 1984, p. 90.

[29] The quotes in this paragraph are taken from Hulme and Sanghi, *Improvement*, p. 4. Cf. also Amin, *Sugar*, p. 56.

> sells at a fairly high price, and this is so because gur has certain peculiar uses, particularly in the Indian culinary art, which have served to maintain the demand even against its formidable rival in the shape of imported sugar from Java, Cuba, and elsewhere.[30]

Despite the fact that imported, white sugar was often cheaper than domestically manufactured *gur* or muscovado, most Indian consumers preferred to stick to the foods they had grown up with. In the beginning of the twentieth century, the whole issue had indeed taken on a religious or political dimension:

> Another reason for the continued existence of the indigenous sugar making industry from gur on a cottage scale is the demand from the orthodox section of the Hindu community for the indigenous stuff even at a price higher than that of the imported commodity.[31]

Sanghi also observed the tendency to prefer domestic products. In the wake of the growing Indian independence movement around the turn of the century, Indian-made brown sugar products became ideologically loaded. If sugar had been produced in accordance with "indigenous" methods, it was what some people "called *swadeshi* sugar for which orthodox Indians will pay higher price than can be obtained for the high class modern factory sugar."[32]

Swadeshi is Hindi, and it may be translated as "homemade." Its invocation in the context of sugar-consumption is logical yet noteworthy. The paradigmatic example of a *swadeshi* cottage industry was the cultivation of cotton and the manufacture of cotton yarn, textiles, and clothes. This was symbolized by Mahatma Gandhi spinning his own yarn at a hand-driven *charkha*. Historians such as C.A. Bayly have discussed domestically manufactured products in terms of their economic, aesthetic, ethical, and political implications, especially compared to imported—mainly British—commodities.[33] To protect domestic crafts, representatives of the *swadeshi* movement tried to convince contemporaries to renounce foreign

[30] Datta and Basu, "Improvements," p. 3.

[31] Ibid.

[32] Hulme and Sanghi, *The Improvement*, p. 5.

[33] Bayly, C.A., "The Origins of Swadeshi (Home Industry): Cloth and Indian Society, 1700–1930," in: Appadurai, Arjun, ed., *The Social Life of Things: Commodities in Cultural Perspective*. Cambridge: Cambridge University Press, 1986: 285–321.

products, even if such a step engendered sacrifice. Bayly quotes a song composed by the nationalist singer Rajanikanto Sen around 1905:

> Defend your homes, protect your shops,
> Don't let the grain from our barns be looted abroad.
> We will eat our own coarse grain and wear the rough, home-spun cloth,
> What do we care for lavender and imported trinkets.
> Foreigners drain away our Mother's milk,
> ...
> We will rather be poor and live our simple lives,
> No one can then rob us of our self-respect.[34]

Rather than buying foreign products, Sen asked his fellow Indians to purchase domestic goods, even if such goods were sometimes more expensive and less attractive than imported items.

The *swadeshi* ideology maintained that domestic items had the virtue of being pure. For example, while domestically manufactured cloth may have been rough to the touch, the argument was that it was better—because it was natural and unpolluted. Conversely, proponents of the *swadeshi* ideology suggested that chemically dyed cloth from Europe was unclean and artificial.

Sugar, too, was drawn into the vortex of the *swadeshi* movement. Newspapers reported that imported sugar contained cow fat, a claim that consumers had no way of evaluating, but a powerful claim nonetheless.[35] Even domestic sugar produced in large factories, under industrial circumstances, came under attack. It was widely believed that labor conditions in sugar factories were inhuman and that bonemeal from cows was used in the production process.[36]

The Improvement Discourse

In the early twentieth century, the *swadeshi* concept thus became an important component of a nationalist ideology. Notably, *swadeshi* was a term applied by Indians themselves, most often in opposition to British products and methods. In contrast, the word *desi* emerged as shorthand to refer to all things considered typically Indian, and it was often used in a

[34] Quoted from Bayly, "Origins," p. 320.
[35] Ibid., pp. 285, 312.
[36] Amin, *Sugar*, p. 104.

derogatory way by British colonizers and foreign observers. Today, the designation *desi* is more neutral and is used by people of Indian as well as Pakistani and Bangladeshi descent.

British interest in so-called *desi* methods of growing sugarcane and making sugar dates to the era when the East India Company tried to turn the subcontinent into a subservient provider of raw materials and semi-finished products. Given the long tradition of sugar manufacturing in India, the company hoped to tap domestic cultivators and producers for their expertise. In 1793, for example, medical doctor W. Roxburgh sent "An account of the Hindoo method of Cultivating the Sugar Cane and manufacturing the Sugar and Jagary" to the company's headquarters in Madras, today's Chennai.[37] Observations about sugar production also abound in Francis Buchanan's three-volume diary of a journey undertaken in 1800 through southern India. Although Buchanan, a physician and botanist, decried "the wretched state of mechanics among the *Hindus*," he nonetheless recorded in minute detail the methods and techniques used in the region.[38]

In these early years of European expansion, British observers and scientists investigated indigenous sugarcane varieties and established sugar-production methods. Just as French biologists scoured West African forests for knowledge and potentially profitable products (please see Chap. 3), British botanists searched the Indian subcontinent. While they recognized the long tradition of sugarcane cultivation and sugar manufacturing in India, their accounts still carried a critical overtone. Like Sanghi, observers and scientists maintained that the *desi* way of operating left much to be desired. Their argument went something like this: to increase productivity and improve the final product, Indian peasants needed to cultivate sugarcane varieties that promised higher yields, and sugar manufacturers needed to invest in machinery that was more efficient. The discourse that developed was one in which "improvement" took center stage.

The improvement discourse derives from reports (1836 and 1837) directed to the British Board of Revenue in Madras. The author of the reports is Robert Wight, a Scottish surgeon and botanist who worked for

[37] Roxburgh's report is part of a file containing several documents from the years 1792–1794 in the India Office Records at the British Library: "Madras. Public Consultations, 29th Mar. to 2nd June, 1794," File No. IOR/P/241/46.

[38] Buchanan, Francis, *A Journey from Madras through the Countries of Mysore, Canara, and Malabar*, 3 Vols. London: Bulmer and Co., 1807; here: Vol. 1, p. 342.

the East India Company. Like Roxburgh's accounts, Wight's reports were "prepared ... for improving the culture of commercial or Mercantile produce in India." One of Wight's suggestions was to introduce sugarcane varieties from Mauritius; supposedly, this would allow farmers to grow cane without extensive irrigation.[39] Following this advice, the Revenue Department promptly "applied to the Mauritius Government for a supply of Cuttings." The choice of Mauritius, instead of a locale like the Dutch East Indies, was likely based on the fact that Mauritius was a British colony at the time.

In his first report, Wight begins by reproducing the standard narrative of the complacent Indian farmer, whose methods were purportedly too primitive and whose products were unsuitable "for the supply of foreign markets":

> The apathy, not to say aversion, of the natives to all improvements suggested by the Government or emanating from its Revenue Officers is proverbial, and is dwelt upon in nearly all the reports, form an almost unsurmountable obstacle to the success of many measures proposed for their benefit.

After these opening remarks, Wight, perhaps surprisingly, suggests that the British themselves develop a deeper understanding of the situation: "we must not, however, it appears to me, lay the whole blame on the cultivator." Wight asserts that the real problem is not the poor peasant himself, but the economic system as a whole: during years of bad harvest, the peasant has no alternative but to borrow money "at exorbitant interest," leaving him "destitute." Aptly, Wight concludes that it "is not to be expected that men so situated would willingly depart from the beaten track and enter on a new and untried field of adventure."[40]

Despite these insights, Wight still believed it would be possible to change the rural system slowly but surely. Just as Sanghi would do eighty years later, Wight put his faith in individual investors and larger landlords, people with the resources required to introduce new crops and new technologies. To support these adventurous pioneers, Wight suggests devising

[39] Information and quotations in this paragraph are taken from the India Office Records at the British Library: "Proposals of Dr. Robert Wight for improving agricultural production in India," File No. IOR/F/4/1695/68233.

[40] The quotes in this paragraph are taken from the India Office Records at the British Library: "Proposals of Dr Robert Wight for improving agricultural production in India," File No. IOR/F/4/1695/68233.

"a course of experiments" to test both Mauritian sugarcane and indigenous "countrycane," noting, "[i]f the countrycane can by a progressive change in the mode of culture be accommodated to an altered system of management, it seems not unprobable [*sic*] that its quality will be improved."[41] According to the correspondence with the Madras Board, Wight even undertook experiments of his own to design: "an improved sugar bailing apparatus and a light plough suited to this Country but on the best English models."

Archival sources indicate that the Madras Revenue Department continued to express interest in trials with imported sugarcane varieties from Mauritius.[42] The authorities also went on to support agricultural technology more generally. In 1867 and 1868, for instance, the Secretary of the Madras Government summarized experiments that had recently been carried out with "English Agricultural Implements." In line with what Wight had suggested, the British authorities asked landlords, factory managers, and even maharajas to test plows, hoes, scythes, and other tools. Their conclusions were in sync with those later made by Sanghi. "English Ploughs" proved four times more efficient than a "Country Plough," but "the drawback is that the ordinary kind of cattle of this country are not strong enough to drag them." In addition, the higher costs of imported British equipment—or equipment made in India in accordance with "English patents"—were prohibitive for most buyers. The Maharajah of Vizianagaram argued in a report to the regional British tax collector that, if the English "implements are adapted in shape, durability, and cost, to the climate, present wants, and conditions of the people, they will find a better demand than they do now."[43] Indeed, the imported equipment was inappropriate in most Indian environments. The contemporary critic A.O. Hume made the ironic comment that those who ventured to use British steam plows in India "seem to have about as accurate a conception

[41] This and the previous quotes in this paragraph are taken from the India Office Records at the British Library: "Proposals of Dr Robert Wight for improving agricultural production in India," File No. IOR/F/4/1695/68233.

[42] Cf. the India Office Records at the British Library: "Endeavours of Sakaram Narayanji Kirulkar, Jagirdar of Manje Malandi in the Deccan, to produce a superior type of sugar from Mauritius sugar cane," File No. IOR/F/4/1907/81545.

[43] The quotes in this paragraph are taken from the India Office Records at the British Library: "Collections to Revenue Despatches to Madras (1868)," File No. IOR/L/E/3/754.

as a certain Maharajah" who wanted to send an Indian elephant to a physically disabled Londoner.[44]

In the last third of the nineteenth century, the British practice of encouraging landlords and peasants to try out new crops and techniques intensified. State governments throughout India began to establish specially equipped farms to demonstrate the efficiency of new equipment; improved fertilizers, and what they claimed were better plant varieties.[45] To guarantee stable test conditions and simplify documentation, the State governments also founded agricultural research stations and production test sites. Sugarcane cultivation and sugar production became important topics of inquiry at several of these institutions. Hulme's Government Experimental Sugar Factory in Uttar Pradesh was one of these many institutions.

As far as possible, experiments at government farms, factories, and research stations followed established scientific protocols. When analyzing the chemical composition of sugarcane juice at the Lyallpur (Faisalabad in today's Pakistan) Farm in the Punjab, for example, the chemists compared juice produced by the "Laboratory Hand Mill" to the juice from the "Farm Mill."[46] Other variables were the variety of sugarcane and the kind of fertilizer used.

Experiments were carried out on the premises of government farms and research stations as well as in the field. One method was to provide peasants in the region with, for example, new sugarcane varieties and to have "out-door assistants" collect information and data from local plots. In this way, ordinary *ryots* (peasants) became research assistants. Surprisingly, the researchers do not treat the peasant as an anonymous, impersonal group, but mention the *ryots*' names. Alladitta, son of Kaka, in village Ghaznipur, and Budda Singh, son of Takhat Singh, in village Dhariwal, were two of the workers. There is evidence, however, that the peasants disliked their role as field researchers and tried to avoid delivering exact data to the scientists. Agricultural chemist J.H. Barnes had to admit that it was virtually "impossible to obtain any reliable figure of the yield of cane per acre, all enquiries in this direction being received with suspicion by the

[44] Hume, A.O., *Agricultural Reform in India* (1879), quoted from Whitcombe, Elizabeth, *Agrarian Conditions in Northern India, Vol. 1: The United Provinces under British Rule, 1860–1900*. Berkeley, CA: University of California Press, 1972, p. 108.

[45] For the United Provinces, see Whitcombe, *Agrarian Conditions*, pp. 103–104.

[46] Barnes, J.H., *Progress Report for 1913 on Sugar in the Punjab*. Lahore: Punjab Government Press, 1913, Table 24.

cultivators."[47] Considering the enormous pressure put on the peasants by landowners, moneylenders, and tax collectors, this suspicion is understandable.[48] J.B. Fuller, Director of Agriculture in the Central Provinces, was well aware of the problem:

> The agricultural classes are notoriously suspicious of any attempts of Government to collect information on matters concerning them, and commonly believe that increased knowledge will certainly result in increased taxation.[49]

Referencing the sugar-production figures he presented to the Chief Commissioner of the Provinces, Fuller admitted that "much reliance cannot be placed on their accuracy," and that some of the numbers boiled down to "mere guess work." Thus from the point of view of the authorities, it made sense to create research stations where experiments could be carried out under circumstances far easier to control.

As mentioned, the immediate impact of demonstration farms, experimental factories, and research stations on agricultural practices in India was limited. One reason was the marked difference between these government institutions and the realities in the countryside. G.H.M. Ricketts, Government Commissioner in the North-Western Provinces in the early 1870s, criticizes the unrealistic investigations carried out at the Cawnpore Experimental Farm:

> No native farmer, however well off he may be, or however enterprising, can hope in his own land to attain anything like the success that has been attained occasionally in experiments conducted under this farm. He sees at a glance that the conditions of the farm and his own fields are so different, that the results attained in the farm are beyond his reach. He sees there ... an unlimited supply of water, manure, costly implements, machinery, labour and supervision, and capital.[50]

[47] This and the former quote come from Barnes, J.H., *Sugarcane in the Gurdaspur District*. Lahore: Punjab Government Press, 1912, p. 7.

[48] The precarious plight of peasants in the Punjab is described by Banerjee, Himadri, *Agrarian Society of the Punjab (1849–1901)*. New Delhi: Manohar, 1982

[49] Fuller, J.B., "Sugar Production in the Central Province," Report dated 23 August, 1883; in the India Office Records at the British Library: File No. IOR/V/27/515/27.

[50] Report by G.H.M. Ricketts to the Government of the North-Western Provinces (1873); quoted from Whitcombe, *Agrarian Conditions*, p. 104.

Piecemeal Change

Criticism like Ricketts' motivated some researchers to rethink their approach. Instead of trying to convince *ryots* to implement the most up-to-date scientific insights or to invest in expensive British machinery, researchers became increasingly open to encouraging peasants to make minor modifications in their traditional methods. Three decades after Ricketts denounced the work at the Cawnpore Farm, W.H. Moreland, Director of Land Records and Agriculture in the United Provinces (which had replaced the North-Western Provinces in 1902), and his assistant, S.M. Hadi, realized the need to avoid "using appliances requiring a disproportionate capital outlay, or introducing chemical processes that would require scientific knowledge …"[51] Among the simple and cheap low-tech solutions Hadi recommended was the adoption of "kerosene oil tins now available everywhere in India" to gather the sugarcane juice extracted by the crushing mill. Such tins were meant to replace "large earthen vessels permanently sunk in the ground," which were, as a rule, both too big and notoriously difficult to keep clean.

Hadi is an interesting figure. His full name was Saiyid (Sayd) Mohammad Hadi, and as his name indicates, he was Muslim. Hadi had the rare opportunity (for an Indian and a Muslim) of studying at the Royal Agricultural College in Cirencester, England. He joined the Agriculture Department of the North-Western Provinces in the 1890s and soon advanced to the position of Assistant Director of Land Records and Agriculture. For several decades, Hadi remained one of the state's most important researchers, spending most of this time working at the Cawnpore Farm. Early in his career, he became a member of the Royal Agricultural Society in London and later he was awarded the prestigious title *Khan Bahadur*, "brave leader," by the British Indian government. In the late 1920s he took on assignments with the Bhopal State in central India.[52] Throughout his active life he spent "much time, thought and energy to the improvement of indigenous processes" of sugarcane cultivation and sugar

[51] Moreland, W.H., "Introduction," in: Hadi, S.M., *Improvements in Native Methods of Sugar Manufacture*. Agricultural Series, Bulletin No. 19, Department of Land Records and Agriculture, United Provinces of Agra and Oudh. Allahabad: Government Press, 1905: 1–3; here: 1, 3.

[52] These brief biographical data are taken from Hadi, Khan Bahadur S.M., *The Indian Sugar Industry*. Bhopal: Government of Bhopal, 1929, title page.

manufacturing.[53] Rather than trying to introduce modern machinery and scientific methods, Hadi took domestic, Indian techniques as his point of departure.

A monograph published by Hadi in 1902 reveals this bottom-up approach. His discussion of sugarcane varieties is telling. Experiments from various research stations and experience from other parts of the world suggested that the type of sugarcane grown in Northern India for centuries should be replaced by other varieties. Although many foreign varieties may well have yielded more juice than the locally grown variety called *sarauti*, Hadi was, nevertheless, not prepared to recommend discontinuing use of the domestic plant, stating, "it gives very satisfactory results under ordinary inexpensive methods of cultivation and with less care than other canes demand." The same went for *ukh*, one of the most widespread sugarcane varieties in the United Provinces. In the harsh climate of Northern India, *ukh* had the big advantage of being "less susceptible to disease and injurious effects of weather on account of their hardy nature." The fact that *ukh* also had been found "less liable to injury from attacks of wild animals" than other sugarcanes showed that the traditional varieties were well-suited to local conditions. Just as Sanghi had concluded that indigenous muscovado-production technologies were well-matched to the local economic and social structures, Hadi claimed that *sarauti* and *ukh* were well-matched to the local climatic conditions. Although other technologies and sugarcane varieties might have been more efficient from a scientific point of view, I suggest that the *rab bel*, *ukh*, and *sarauti* were *appropriate* to the local context of Northern India.[54] They were, so to speak, appropriate technologies.

Hadi's approach was a hybrid one. He exhibited an astute understanding of local needs and circumstances, and he did not want to recommend varieties and methods on abstract, theoretical grounds. At the same time, he was a scientist, and as such, he gave in to the temptation to carry out "experiments in different parts of the country under our personal supervision." When analyzing the "outturn" of various sugarcane plants, Hadi compared as many as sixty-seven different varieties, measuring how much

[53] Richards, P.B. "Foreword," in: Hadi, Khan Bahadur Syed Mohammad, *Rab, Khand, and Gur Making with Hadi Miniature Bel.* Bulletin No. 74, Department of Agriculture, United Provinces. Allahabad: Superintendent, 1938: i.

[54] The quotations in this paragraph are taken from Hadi, Saiyid Muhammad, *The Sugar Industry of the United Provinces of Agra and Oudh.* Allahabad: Government Press, 1902, pp. 3–4, 8.

the harvested cane weighed per acre, how much juice could be extracted, and how much *gur* could be produced from a specified amount of sugarcane. Although he tried to be as systematic and as comprehensive as possible, he considered his results to be merely preliminary. The experiments were carried out at various sites, where different technologies were deployed. Some locations had iron mills while others had wooden mills; some mills had three rollers, while others had only two. In addition, each plot and each tract had its own characteristics—including distinct climates—which made direct comparisons inconclusive. To make things worse, the hydrometer Hadi and his team had used to measure the amount of sugar in the juice proved unreliable.[55]

The hybrid character of Hadi's methodology is evidenced by two facts: he did not rely exclusively on established scientific knowledge, nor did he use only standard chemical procedures. Hadi took cultural factors and local expert knowledge seriously; in his 1902 monograph, he even included a chapter on "Ceremonies and Superstitions Connected with the Cultivation of Sugarcane." In the process of finding a suitable method to clarify sugarcane juice, Hadi and his colleagues adopted a "clarifying mixture recommended by a professional boiler of Ahábjahánpur." After experimenting with this mixture, the Agricultural Department concluded that "it imparts a beautiful colour to the resulting *gur*, and thus enhances its value in the eye of the purchaser requiring it for consumption as food." This case is also interesting in that it reveals Hadi's urge to adopt domestic, vernacular expressions. Among other things, the professional boiler's recipe included *gáozabán* (from the bugloss plant) and *káli sajji* (sodium carbonate or washing soda).[56]

Hadi was convinced that the best strategy for protecting domestic sugar production was to modify existing technologies by degrees. Unlike many colonial economic policymakers who argued in favor of large-scale, centralized factories, Hadi trusted the small *ryots* and the local manufacturers to make necessary changes. In the late 1920s, when referring to the situation in the state of Bhopal, Hadi concluded with satisfaction that the "old stone mills have now been completely replaced by the iron mills, though a large number of the former still exist … as relics to recall to the cultivator's mind how his forefathers dealt with their cane harvest in the old days." Further data indicated that the three-roller mill had "gained almost

[55] Hadi, *Sugar Industry*, pp. 28–36.

[56] The quotes in this paragraph are taken from ibid., pp. 63, 76.

universal approbation from the cane-growers of the United Provinces and the Punjab."[57] Hadi suggested that change was possible—without turning the existing production system inside out.

Contrary to their alleged resistance to change, Indian peasants did, in other words, adopt new technologies. Detailed statistics from as early as 1892 reveal that one-third of the cane-growing districts in the Madras Presidency in Southern India had already acquired portable cane-crushing mills made of iron. What made these portable mills appropriate was the fact that they were light enough to obviate the need for multiple—or stronger—steers.[58]

The adoption of new technologies differed from region to region. The Thompson and Mylne Company marketed the Behia cane-crushing mill with some success in Bihar, the United Provinces, and the Punjab. Compared to the traditional mill, the Behia mill (named after an estate in Bihar) proved to be approximately thirty percent more efficient. Given that it was easier to handle, the Behia mill also required less labor than traditional wooden mills. For the low-income *ryot*, however, the decisive factor was not the machine's greater efficiency but the question of whether or not the Behia mill required further investment. Fortunately, the mill did not require the peasant to acquire more—or stronger—steers.[59]

Promoters of the Behia mill were less successful in the Central Provinces and the Bhopal State. In the early 1880s, despite the efforts of some of the government's deputy commissioners, few peasants in the Central Provinces were convinced to purchase Thompson & Mylne's products. Director Fuller explains the sound economic logic behind this reluctance. In the context of rural India, the argument that the Behia mill required less labor than the traditional mill did not carry much weight. In this environment, where labor costs were extremely low and most work was done by family members, why would anyone invest money in labor-saving equipment—or, worse, borrow money to do so? After all, the Behia mill cost roughly forty times more than a wooden mill. Just as important, wooden mills were "made by the village carpenter," who could also repair them when necessary. Although his position on the social ladder was fairly low, the

[57] The quotes in this paragraph are taken from ibid., p. 87.

[58] Benson, C., *Sugar Mills*, Department of Land Records and Agriculture, Madras, Agricultural Branch: Bulletin No. 26. Madras: Government Press, 1892.

[59] Remark by Wood in *Extracts from the Revenue Administration Reports of the Punjab 1884–1885*, quoted from Banerjee, *Agrarian Society*, p. 64.

carpenter carried out important tasks and belonged to the village's social structure. He was a person who "the village could not well dispense with."[60] Fuller's observation is exciting in that it highlights the embeddedness of technologies in the social landscape. If peasants chose to buy iron mills, they ran the risk of losing their carpenter.

Factory Sugar

Most agricultural researchers and civil servants likely disagreed with Hadi's prudent position. The standard view in expert circles was that sugarcane cultivation and sugar production needed radical reform: in the face of foreign competition, peasants should adopt high-yielding sugarcane varieties, and production should be carried out on a large scale. The vision was clear: directors of demonstration farms, experimental factories, and research stations wanted higher investments, bigger output, and greater efficiency. Centrally placed factories belonged to this vision. Given that they were highly mechanized, such factories would take in sugarcane from a fairly large area; if necessary, the sugarcane would be transported to the factory by rail. When J.H. Barnes described the research and experiments undertaken at the government's farms in the Punjab, he made it abundantly clear that:

> one of the main objects [*sic*] of the enquiry was to ascertain the possibility of obtaining canes, good enough in quality and high enough in yield per acre to offer an opportunity of successfully establishing a centre factory for the production of gur and refined sugar.[61]

The purpose of much research and development work was to pave the way for the industrialization of the sugar business.

From a macroeconomic perspective, the experts' modernization policy made sense. Throughout this period, sugar producers in India faced the constant, looming threat of their products being squeezed out of the market by cheaper, refined cane sugar from Java. The introduction of the sugar beet in Europe had only made the situation worse. In the second half of the 1890s, Hadi himself had expressed a fear that beet sugar could displace domestic sugarcane: "even the poorest opium smoker of Lucknow

[60] The information and quotes in this paragraph come from Fuller, "Sugar."
[61] Barnes, *Sugarcane*, p. 7.

could afford to indulge in the luxury of sweetening his tea with the brilliant foreign stuff—a pleasure he had never dreamt of before."[62] As mentioned, Hadi's fears would prove somewhat exaggerated. Refined sugar did make inroads into some urban areas, but the majority of Indian consumers continued to prefer *gur* and muscovado.

Nevertheless, the constant threat of competition from the global market continued to worry governments and committees. In 1919 the Governor General appointed the Indian Sugar Committee to investigate "the possibility of organising and developing the Sugar Industry in India." Two years later, under the chairmanship of James MacKenna, Director of the Agricultural Research Institute in Pusa, the committee published a strikingly comprehensive report of almost five-hundred pages. Over the course of more than eighty meetings, the committee had discussed technical and economic matters along with legal, agricultural, and organizational concerns. Virtually all their recommendations were in service to saving the Indian sugar business from collapse; to achieve this, the entire production process would need to become far more efficient. MacKenna and his colleagues called for wide-ranging structural "improvements": the traditional "systems of land tenure" had to undergo reform; small-scale *gur* manufacturers needed to reduce waste and introduce "less costly and more efficient methods" as far as possible, and "large modern factories" were to replace the existing "cottage industry."[63]

The Sugar Committee argued in favor of two kinds of factories. On the one hand, committee members were convinced that, in the long run, factories producing white, refined sugar would make sugar production in India more competitive; if successful, these factories could contribute to turning India into a net exporter of sugar. On the other hand, the committee realized that manufacturers needed to continue to meet the domestic demand for *gur* and muscovado. MacKenna's final report included one chapter on "The Sugar Factory and Its Raw Material" as well as a chapter on "The Manufacture of Gur." The first of these chapters is interesting in that it mainly concerns "the supply problem." The report writers reasoned that even the most modern, highly mechanized industrial plant would become profitable only if a sufficient supply of sugarcane from the surrounding tracts could be guaranteed. In the case of *gur* production, the

[62] Hadi, *Sugar Industry*, Preface.

[63] The quotes in this paragraph are taken from *Report of the Indian Sugar Committee*. Simla: Government Central Press, 1921, pp. 1–2, 261, 268.

committee recommended installing steam-powered cane-crushing mills and "hand-driven centrifugal machines."[64]

MacKenna's committee engaged in the "improvement discourse" seen also in Wight's writings of the 1830s. Given that one percent of all sugar products in India were "made directly in the form of factory sugar," the transformation of the trade would be a formidable task. The cultivation of sugarcane had to be expanded to—and intensified in—areas surrounding sugar factories, by leasing land to farmers who specialized in sugarcane, if necessary. According to the committee, the manufacturing process, start to finish, needed to be modernized. Factory owners would have to invest in "labour saving appliances" and hire scientifically trained personnel: "The first essential for improvement is better supervision and above all more efficient chemical control." To show potential investors how a modern factory should be run, MacKenna recommended that they commission "a pioneer factory" to be built in the Punjab.[65]

MacKenna's recommendation raises the question of why the Indian Sugar Committee would seek to build a new demonstration factory, when Hulme's Government Experimental Sugar Factory, in the United Provinces, was already available as a prototype. Indeed, the committee members discussed Hulme's Nawabganj factory at great length, but they criticized heavily some of Hulme's methods. Whereas MacKenna praised the eleven-roller crushing mill for being "capable of giving a very high extraction," he did not agree with the decision to boil the sugarcane juice in open pans rather than in high-pressure vacuum pans. The committee indicated that boiling the juice in open pans, at atmospheric pressure, would always produce detrimental results.[66]

The Sugar Committee went so far as to suggest closing down the Nawabganj factory. The outright denunciation of Hulme's research-and-development strategy demonstrates a certain tension among the experts. For their part, Hulme and Sanghi had designed a factory that was supposed to be "the smallest plant that could be worked on a commercial basis." Their decision not to install the purportedly more efficient vacuum pans is telling: they described the vacuum-pan apparatus as far too costly and as requiring "so much skill to manipulate."[67] Hulme's goal had been

[64] The quotes in this paragraph are taken from ibid., pp. 265, 276, 281, 291, 321.

[65] The quotes in this paragraph are taken from ibid., pp. 261, 339.

[66] The quote in this paragraph is taken from ibid., pp. 270.

[67] This and the previous quotes are taken from Hulme and Sanghi, *Note*, p. 2.

to "improve indigenous methods" and to design a factory that could be operated by employees without advanced scientific and technological training. The government of the United Provinces had first become aware of Hulme's ideas in 1911, and three years later the Experimental Factory had been set up. In 1921, the Sugar Committee declared Hulme's attempts to be an outright "failure."[68]

Intermediate Technology

The tensions between various factions in India's sugar business were clear. At the modernization end of the continuum, the majority of Indian Sugar Committee members subscribed to a typical, ideal-industrialization paradigm. Convinced that the centralization and mechanization of production were absolute necessities, the majority lobbied for the use of scientific knowledge and methods; MacKenna himself wished to see the dawn of "Modern Technology" for India's sugar business.

At the traditionalist end of the continuum, Indian experts like Hadi argued in favor of a low-tech production paradigm—including its advantages and disadvantages. As advantages, this paradigm offered highly affordable means of production; the accessibility of local resources; and the ability of workers to exercise skills that had accumulated throughout the centuries and to continue to serve their community. In addition, manufacturers delivered sugar product that matched domestic consumption patterns. The disadvantage of the system was monolithic: it meant the subjugation of the peasant in an unjust economic and legal system.

Hadi's ideal production system is what I refer to as the paradigm of *appropriate technology*. And somewhere in the middle of the continuum we find the approach of William Hulme and Ram Richh Pal Sanghi, who, according to my interpretation, tried to develop what economist E.F. Schumacher called *intermediate technology*.[69] Their factory required modest investments, and it was meant to make peasants' sugar production competitive on the free market. At the same time, their approach was based on "indigenous methods" and did not depend on highly specialized workers.

[68] *Report*, p. 272.

[69] Schumacher, E.F., *Small is Beautiful: A Study of Economics as if People Mattered*. London: Abacus, 1973; cf. Chap. 7 in this volume.

Schumacher has been called the godfather of the appropriate-technology movement.[70] In his own writings, however, Schumacher most often used the concept of "intermediate technology" to denote small-scale industries, "relatively simple" production methods, and the use of "local materials."[71] Schumacher emphasized that his "idea of intermediate technology does not imply simply a 'going back' in history to methods now out-dated." He argued for the creation of factories rather than a return to preindustrial forms of production. According to Schumacher, such factories should make use of "fairly simple and therefore understandable" equipment, and they should allow for the employment of workers without extensive formal education. Had Schumacher been active half a century earlier, he may well have defined Hulme and Sanghi's Experimental Farm as a typical example of a factory that embraced intermediate technology.

Schumacher was well informed about the situation in the "Third World," as it was then called. In fact, he spent a considerable amount of time in Burma (now Myanmar) and gave talks on Gandhian economics.[72] In his central work, *Small is Beautiful: A Study of Economics as if People Mattered*, Schumacher explicitly refers to Indian experiences when discussing intermediate solutions. In a chapter of *Small is Beautiful* based on a 1965 speech Schumacher delivered in Santiago de Chile, he quotes a passage from a book which the Indian economist Dhananjay Ramchandra Gadgil had published one year earlier:

> One approach may be to start with existing techniques in traditional industry and to utilise knowledge of advanced techniques to transform them suitably. Transformation implies retaining some elements in existing equipment, skills and procedures ... This process of improvement of traditional technology is extremely important, particularly for that part of the transition in which a holding operation preventing added technological unemployment appears necessary.[73]

It is striking how Gadgil's description echoes Hulme, Sanghi, and Hadi's thinking on how to improve "indigenous methods." The traditional

[70] Riedijk, W., ed., *Appropriate Technology for Developing Countries*. Delft: Delft University Press, 1982.

[71] Schumacher, *Small*, p. 165, 176.

[72] Roszak, Theodore, "Introduction," in: Schumacher, *Small*, pp. 1–9.

[73] Gadgil, D.R., *Appropriate Technology for Indian Industry* (1964), quoted from Schumacher, *Small*, p. 177.

methods of cultivating sugarcane and manufacturing *gur* and muscovado were indeed problematic from an ethical standpoint, but they were nonetheless well-suited to India's social, cultural, economic, and environmental realities. For these reasons, indigenous methods can be considered *appropriate*. What Schumacher had in mind when quoting Gadgil is another method, however. It is the method of *intermediate* technology.

Gadgil was a professor at the Gokhale Institute of Politics and Economics in Poona (today's Pune), in the Western part of India. Although likely a case of mere coincidence, it is interesting to note that Poona had been the birthplace of a sugar-making furnace that bore all the characteristics of an intermediate technology: the so-called Poona Furnace. This device was used in the production of *gur* at the beginning of the twentieth century. In a bulletin from 1911, the Divisional Inspector of Agriculture in the Bombay Presidency, P.C. Pa´til, describes the advantages of this furnace. What Pa´til observes is indeed a version of intermediate technology:

> This Bulletin does not profess to explain the most scientific or up-to-date method of evaporating sugarcane juice. It is designed only to show a simple and very economical change which may easily be introduced by cultivators in many parts of the Bombay Presidency, and which involves very little alteration in the present organisation. ... The Poona furnace is built on scientific principles, though of a primitive type.

Compared to the commonly used *gur* furnaces in the region, the Poona Furnace was somewhat more elaborate and required considerable brickwork. Using the Poona Furnace also meant that peasants or *khandsaris* were required to invest in larger boiling pans. The construction of the furnace could be mastered in most villages, though not without prior instruction. As explained by an individual named Pandurangao Anandrao Mohite, in a letter to Pa´til: "Often cultivators cannot construct this new furnace properly, and so I hope the Government officers will keep an eye on this."[74]

Regardless of the tensions between representatives of modern, intermediate, and appropriate solutions, all actors have one thing in common: they worried about what would happen if India were to be pulled into the

[74] The quotations in this paragraph are taken from Pa´til, P.C., *Poona Furnace*, Department of Agriculture, Bombay: Bulletin No. 48. Bombay: Government Central Press, 1911, General Remarks, pp. 1, 9–10.

vortex of global trade—and drowned by foreign competition. These worries proved to be unfounded. Most people in India continued to consume *gur*, muscovado, or domestically refined sugar, and in 1932 the Indian government introduced heavy customs duties on imported sugar. Unlike many histories of sugar, the history of Indian sugar is not a global history of a dynamic commodity; it is a national and regional history of a relatively static product.

PART II

Twentieth-Century Improvisations

CHAPTER 5

Accessing Electricity in East Africa: Dar es Salaam Dwellers Pursue Power

RE: INSTALLATION OF A LAMP POST NEAR THE HINDU MANDAL PREMISES.

With reference to your letter No. LT/1/202 of the 24th ultimo, I shall be obliged if you will write to me as to when the funds are going to be available and whether the same have been applied for at all or not. There is no indication whatsoever regarding the same in your letter under reply.

The installation of a lamp post near the Hindu Mandal premises is absolutely necessary and urgent and requires special consideration in view of the fact that the Hindu Public Dispensary being in that very building, both the Doctor and the people experience great difficulty while going there late at night.

An early reply is solicited as, in case no satisfactory reply is received from you, further communication with higher authorities is to be made immediately.[1]

Apparently, Mr. C. Patel, Honorary Secretary of the Hindu Public Dispensary in Dar es Salaam, was losing his patience. Despite several efforts to persuade the Public Works Department (PWD) to install an electric lamppost in the narrow Kisutu Middle Street that led to the Hindu hospital, nothing had happened. The hospital treated up to seventy patients per day, and those who arrived after dark had difficulty finding

[1] Letter by C. Patel to the Director of Public Works, March 3, 1933; in: Tanzania National Archives (hereafter: TNA), File No. 18878, "Street Lighting Dar es Salaam."

M. Hård, *Microhistories of Technology*, Palgrave Studies in the History of Science and Technology,
https://doi.org/10.1007/978-3-031-22813-1_5

their way. The medical doctors were also bothered by the "inconvenience" of having to grope their way from the main roads to the dispensary after dusk and before dawn.[2] The dispensary—now called the Shree Hindu Mandal Hospital—had been founded in 1919; at the time of Patel's letter, it was fourteen years later, and Patel and his colleagues were still waiting for the PWD to install electric streetlights.

Mr. W.H. McLuckie, the Director of Public Works, repeatedly pointed out that, due to a limited budget, he was unable to meet Patel's request. In an effort to forestall the hospital from appealing to "higher authorities," McLuckie dashed off a letter to the Chief Secretary to the British Governor of Tanganyika Territory, saying, "it is not apparent that the lighting required by the Hindu Dispensary is such that should ordinarily be provided by street lighting and not by an entrance lamp of their own over their premises." McLuckie stressed that "no funds have been provided in the current year's estimates for additional street lighting."[3] Agreeing with McLuckie, the Chief Secretary repeated the official policy: "that Government provides lighting for the purpose of illuminating the public thoroughfare and that it cannot undertake to furnish lamps for the purpose of facilitating access to any particular building."[4]

Fortunately for the hospital, the press intervened. In late May and early June 1933, both the *Tanganyika Herald* and the *Tanganyika Opinion* picked up the story. The *Herald* took a particularly harsh stance toward the government. Under the headline "Scandal," the paper juxtaposed the urgent needs of the hospital and the interests of the well-heeled inhabitants in two newly developed seafront neighborhoods: "Nearly a thousand pounds are being spent in fitting the houses at Oyster Bay and Sea View with electric light, yet a medical dispensary must go without light and road which can be provided at the cost of a few hundred Shillings." At the end of the short article, the journalist urged "His Excellency the Acting Governor ... to pay a visit to this Dispensary and see for himself the

[2] Letter by C. Patel to the Chief Secretary to the Governor, July 14, 1933; in: TNA, File No. 18878, "Street Lighting Dar es Salaam." The Kisutu Middle Street is nowadays called Chusi Street.

[3] Both quotes are taken from a letter by the Director of Public Works to the Chief Secretary to the Governor, March 9, 1933, in: TNA, File No. 18878, "Street Lighting Dar es Salaam."

[4] Letter by the Chief Secretary to the Governor to the Director of Public Works, March 15, 1933, in: TNA, File No. 18878, "Street Lighting Dar es Salaam."

disgraceful state of affairs ... and give immediate orders for the construction of a road and the installation of the electric light."[5]

The newspapers' interventions yielded immediate effects. Although we do not know if Governor Stewart Symes followed the *Herald*'s advice to visit the dispensary, his Chief Secretary wrote to the Director of Public Works a mere three days after the article was published: "I am directed to inform you that on further consideration it has been decided that it is desirable to provide the additional lighting required for the Hindu Public Dispensary." Nevertheless, the governor remained unwilling to commit additional funds to the project. Instead, the Chief Secretary informed the director and the treasurer that the costs were to be covered by "savings from the Road Maintenance Votes."[6]

The case soon made it to the Tanganyika Legislative Council. Obviously unaware that the Chief Secretary had already attended to the matter, Mr. M.P. Chitale—one of the two Council members that were of Indian descent—wanted exact figures: "What is the estimated cost the Government will have to bear for—[*sic*] electrifying Oysterbay and Sea Point View ... What is the estimated burden on public revenue of the light required for the Hindoo Public Dispensary?" Indirectly questioning the government's habit of prioritizing projects in high-end neighborhoods, Chitale asked the governor to consider the fact that the Hindu hospital, by providing medical care to a substantial part of the population, was "indirectly lightening the burden of Sewa Hadji Hospital and so relieving public revenue to that extent."[7]

The PWD finally got its act together. At the end of June 1933, the lamppost was installed.[8] Surprisingly, the story does not end here. Although the streetlight was operational, Patel in mid-July complains that "nothing so far has been done in the matter."[9] Why the complaint? The

[5] "Scandal," *Tanganyika Herald*, May 30, 1933.

[6] Letter by the Chief Secretary to the Governor to the Director of Public Works (cc: the Treasurer and the Auditor), June 2, 1933, in: TNA, File No. 18878, "Street Lighting Dar es Salaam."

[7] Minutes from the Legislative Council, June 9, 1933, in: TNA, File No. 18878, "Street Lighting Dar es Salaam."

[8] Handwritten note on a letter by the Secretary of the Hindu Public Dispensary to the Chief Secretary to the Governor, July 14, 1933, in: TNA, File No. 18878, "Street Lighting Dar es Salaam."

[9] Letter by C. Patel to the Chief Secretary to the Governor, July 14, 1933; in: TNA, File No. 18878, "Street Lighting Dar es Salaam."

answer lies in a letter Patel wrote to the governor's Chief Secretary in early October: "… the lamp which the Government had kindly condescended to provide for the convenience of the patients of the Hindu Public Dispensary has been fixed up at a wrong place."[10] Instead of installing the lamppost in front of the hospital, PWD had placed it at the corner of Kisutu Middle Street and Upanga Road. Apparently, the department continued to follow its principle of lighting thoroughfares and larger roads rather than smaller streets and narrow alleys. Patel himself voiced his suspicion that the lamppost had been placed in that spot because a popular movie-house—the Bharat Cinema—was situated on that very corner.[11]

Who Gets Electricity?

Consider the 1933 correspondence between the Chief Secretary to the Acting Governor of Tanganyika (now Tanzania), a member of the Tanganyika Legislative Council, the Director of the Public Works Department, and the Secretary of the Hindu Public Dispensary in Dar es Salaam. On the face of it, these missives may seem merely bureaucratic; but the letters illustrate the inherent conflict between various interest groups and actors. The correspondence also indicates who was to gain from the expansion of the electricity network in the British-ruled Tanganyika Territory. Constantly constrained by budget restrictions, the governor did everything in his power to contain costs. The Township Authority and its planning staff consistently prioritized the downtown area and the wealthier parts of town where the European minority resided. The representatives of the Indian segment of the population struggled to receive the same treatment as the European dwellers. The Public Works Department tried to please these various groups—as well as to uphold basic engineering standards.[12]

[10] Letter by C. Patel to the Chief Secretary to the Governor, October 2, 1933; in: TNA, File No. 18878, "Street Lighting Dar es Salaam."

[11] When writing this chapter, I profited immensely from knowledgeable comments by Emanuel Lukio Mchome, Jonas van der Straeten, and Frank Edward. They also provided me with some of the material on which the chapter is based.

[12] For a discussion about the difficult position of the Public Works Department in Tanganyika, see Edward, Frank, and Mikael Hård, "Maintaining the Local Empire: The Public Works Department of Dar es Salaam, 1920–1960," *The Journal of Transport History* 41 (1), 2020: 27–46.

Consistently absent in the archival material presented above are the interests of the inhabitants of "native quarters," as they are referred to in the written sources. Since the days of German colonial rule—which lasted from 1885 until the First World War—Dar es Salaam had been divided, on the basis of "race," into three parts. The British referred to these areas as "zones," allocating one zone for "Europeans," who were White; one zone for "Indians and Arabs"; and one zone for "Black Africans."[13] Different rules applied to each part of the city. In Zone 1, for example, real-estate owners were permitted to build only stone houses. This rule effectively banned African people from living in the more expensive neighborhoods, given that virtually no one in the African community could afford to apply such a comparatively expensive construction material as stone. Therefore, this rule amounted to ethnicity and class policy that discriminated against people of African descent in Dar es Salaam. The authorities considered the housing needs of the African community only in terms of their role as interim laborers. Accordingly, planners devoted comparatively few resources to what they disparagingly called "native quarters." Planners considered the dwellings in Zone 3—where the African population of Dar es Salaam lived—to be temporary and substandard. Most of Zone 3 was bereft of streetlights, and African households were not connected to the electricity grid. Only in the late 1930s did the authorities begin to discuss how to provide electricity to the dwellings of "Black African" people. In 1938, the Township Authority launched the so-called Better Lighting Scheme, which included some African neighborhoods; in 1939, the Public Works Department made initial plans to install fifty-three streetlights in the western part of town, where African people dwelled.[14]

Tensions and conflicts around the power grid were played out in the media; on an administrative level; and on the political stage. Supplying electricity had important financial implications, given that the Township Authority paid a yearly fee for each streetlamp *and* financed both the

[13] Vorlaufer, Karl, *Koloniale und nachkoloniale Stadtplanung in Dar es Salaam: Gesellschaftspolitische Zielvorstellungen und städtebauliche Ideen in ihrem Einfluß auf die Raumstruktur einer tropischen Großstadt.* Frankfurt am Main: Frankfurt University, 1970.

[14] Mchome, Emanuel Lukio, *"Blackout Blues": A Socio-cultural History of Vulnerable Electricity Networks and Resilient Users in Dar es Salaam, 1920–2020*, Darmstadt: Technical University of Darmstadt, 2022. https://tuprints.ulb.tu-darmstadt.de/id/eprint/21803 [accessed Aug. 30, 2022], p. 39. See, also, "Plan shewing [sic] street lighting" (1939), in: TNA, File No. 18878, "Street Lighting Dar es Salaam."

maintenance and the expansion of the electric-streetlight network.[15] In the colonial era, authorities did not consider electric lighting to be a necessary utility for everyone in society; authorities excluded those living in Zone 3, for example. The colonial government and the Dar es Salaam Township Authority regarded electricity as a service to be provided for White residential areas, the Central Business District, and the several manufacturing companies in town. Importantly, the British authorities expected the Electricity Department—a subdivision of the Public Works Department—to contain costs and stay within its budget. This harmonized with the British policy—in effect until the 1940s—that the colonies should be as self-sufficient as possible, rather than depend on financial support from London.

Another factor limited the authorities' influence: the Public Works Department itself did not produce the electricity. As we will soon see, power production in the Dar es Salaam area had been in private hands since 1931. Although the Electricity Department financed street lighting, the Dar es Salaam and District Electric Supply Company (DARESCO) was responsible for power generation and providing private customers with electricity. DARESCO was a subdivision of the East African Power and Lighting Company (EAP&L), a Nairobi-based enterprise financed primarily by British capital. DARESCO's business strategy was opportunistic: to supply electricity to parts of the city where the most profit could be made.

In colonial Dar es Salaam, then, electricity provision was a biased process, in which power was dispensed according to class and "race." Although contemporary commentators referred to DARESCO as a "public utility," this appears to be a misnomer.[16]

Whereas electricity provision to private establishments was to a large extent a commercial endeavor, the provision of street lighting constituted a public initiative, and carried strong political implications. As the Hindu Dispensary case illustrates, in the 1930s electric streetlights were a topic of public concern. No one could deny that electricity provision was a matter of both economics *and* politics.

[15] Minutes by a PWD "Electrical Engineer," Feb. 22, 1938; in: TNA, File No. 18878, "Street Lighting Dar es Salaam."

[16] Letter by E.B.B. Speed to J.A. Calder of the Colonial Office, Feb. 14, 1936; in: National Archives, Kew, UK (hereafter: NAK), File: "T 161/1049: Tanganyika: Dar-es-Salaam Electricity Supply"; Jonas van der Straeten was kind enough to provide me with scans of this file.

In his comprehensive study of the electrification of East Africa, scholar Jonas van der Straeten shows that throughout most of the British era, the power sector was a privately held, commercial undertaking. Like other electricity providers in Tanganyika and Kenya, DARESCO did not view its primary responsibility as serving the public. Although it acted within a strong regulatory framework, the company was beholden to its shareholders, most of whom were British residents.[17]

As scholar Rebecca H. Ghanadan discusses in her study of the postcolonial era, the tension between electricity as a "public service" and electricity as a "commodity good" continued in postcolonial Tanzania.[18] Despite attempts in the 1990s to privatize and deregulate the business—which had been nationalized shortly after independence in the 1960s—electricity remained a political issue, Ghanadan argues. Unsurprisingly, electric light and power never became purely commercial commodities that hewed strictly to the laws of capitalism: access to electricity remained a partially sociopolitical issue. While the Tanzanian government required providers to keep electricity prices low, the reform process of the 1990s made it difficult for "disadvantaged populations and groups" to secure contracts with TANESCO (Tanzania Electric Supply Company Limited), Ghanadan shows.[19] Despite being under government control, the company showed evidence of being motivated by profits. Or, as historian Emanuel Lukio Mchome frames it, TANESCO "became more commercial, no longer operating as an agent of economic modernisation as had been the case under the socialist government from the 1960s to 1980s."[20] Persuaded by Ghanadan's argument and evidence, I illustrate in this chapter how electricity provision oscillated between public service and economic commodity—not only in independent Tanzania, but in the British-controlled Tanganyika Territory as well as in German East Africa.

Despite the fact that some historical actors referred to DARESCO as a public utility, I challenge the idea that the organization played this role. It is accurate that the Public Works Department and its Electricity Department were public institutions in the sense that they were under the control of

[17] Straeten, Jonas van der, *Capital Grids: A Global History of Electricity in East Africa*. New York: Palgrave Macmillan, forthcoming.

[18] Ghanadan, Rebecca H., "Public Service or Commodity Goods? Electricity Reforms, Access, and the Politics of Development in Tanzania." *PhD Thesis*, University of California at Berkeley, 2008.

[19] Ghanadan, "Public Service," p. 84.

[20] Mchome, *"Blackout Blues,"* p. 193.

the Dar es Salaam Township Authority. Nevertheless, these institutions were unable to provide all the city's people with equal access to services. Reflecting the political power relations in the city, the Department prioritized lighting streets in the downtown area, in the upper-class neighborhoods, and along larger thoroughfares. In the case of DARESCO, the situation is even more obvious: this was a private company whose primary goal was to remunerate its owners. Deeming DARESCO a "public utility" is, at best, a euphemism. In the case of DARESCO, I consider "service provider" to be a more appropriate term. "Service provider" connotes an institution that gives customers access to a particular resource by means of technical infrastructure—be it energy, water, or data. Suppliers of electricity make use of power lines; water and gas providers lay pipes; and telecommunications companies enable users to communicate wirelessly, via underground cables, and via overhead wires. The fact that "service provider" sounds like a neutral concept should not obscure the political nature of infrastructure provision.

Historians of technology and political scientists have tended to acknowledge the fact that infrastructure services are often subject to political controversy.[21] In his *Networks of Power*, historian Thomas P. Hughes analyzes at length how the early provision of electricity in London, Berlin, and Chicago reflected technological, political, and economic circumstances.[22] On the basis of this analysis, Hughes developed an evolutionary model to describe the early phase and the ensuing expansion of large technological systems. Once a system has established itself commercially, socially, and politically, the thinking goes, the system continues to grow and develop more or less autonomously. Hughes describes this self-governed growth via the physical metaphor of momentum.[23]

Hughes grounded his model in historical analyses of electricity networks in Northern Europe and North America, and most historians of technology have, hitherto, applied Hughes' model to cities in the Global North. But does his model remain relevant when applied to infrastructure in the Global *South*? The Hindu Dispensary case indicates that electricity

[21] Winner, Langdon, "Do Artifacts Have Politics?" *Daedalus* 109, 1980: 121–136.

[22] Hughes, Thomas P., *Networks of Power: Electrification in Western Society, 1880–1930*. Baltimore, MD: Johns Hopkins University Press, 1983.

[23] Hughes gives a summary of his concepts and model in: Hughes, Thomas P., "The Evolution of Large Technological Systems," in: Bijker, Wiebe E., Thomas P. Hughes, and Trevor J. Pinch, eds, *The Social Construction of Technological Systems: New Direction in the Sociology and History of Technology*. Cambridge, MA: MIT Press, 1987: 51–82.

networks in British East Africa in general—and in Dar es Salaam in particular—did not operate according to the logic of Hughes' evolutionary model. As Mchome concludes, the system failed to acquire momentum—either in the interwar period or in the postwar years.[24] Rather, growth took hold in selected parts of the city only. In fact, the spread of electricity remained a contested, highly political process; the level of investment remained low; and demand was sluggish. The electricity networks in Tanganyika Territory remained limited in size, and they were barely interconnected. In fact, most electricity networks operated in "island mode": there were few integrated grids and few large hydropower plants. The piecemeal nature and limited reach of the electricity network in Dar es Salaam constitute evidence of van der Straeten's argument that it is "difficult to apply Hughes' interpretive framework to Africa."[25]

To some extent, however, Hughes' model is helpful to historians of infrastructure in colonial settings. As we have seen, in Dar es Salaam electricity provision was an economic and political issue as well as a technological one. Mchome argues that, compared to the city's European and Indian populations, inhabitants in the "native quarters" were denied fuller access to electricity primarily on the basis of their low income—as well as discriminatory regulations against "non-permanent" buildings. In addition, DARESCO faced technical problems and lacked capital for larger investments.[26]

Similarly, in his history of water provision in Nairobi, Jethron Ayumba Akala argues that the long-standing exclusion of the African population from the centralized water-supply network was based on the fact that the majority of Africans could not afford the necessary infrastructure (pipes) and the water provided by that infrastructure.[27] (Akala posits this in addition to other scholars' theories of racist thinking and "spatial apartheid.") Both Mchome and Akala demonstrate that service provision in colonial East Africa had both an ethnic and an economic component; it concerned both race and class.

[24] Mchome, *"Blackout Blues,"* p. 64.

[25] Straeten, *Capital Grids.*

[26] Mchome, *"Blackout Blues,"* p. 42.

[27] Akala, Jethron Ayumba, *In the Technological Footprints of Urbanity: A Socio-political History of Water and Sanitation in Nairobi, 1899–2015*. Darmstadt: Technical University of Darmstadt, 2019. https://tuprints.ulb.tu-darmstadt.de/8550/ [accessed Aug. 19, 2022], p. 236.

When Electricity Signaled Power

The first electricity-producing plant in Dar es Salaam was built in 1905, during the period of German occupation by the Berlin-based East African Railroad Company (Ostafrikanische Eisenbahngesellschaft). The plant's initial purpose was to serve the company's own needs. Two steam-powered generators and one battery were installed to drive the machines in the company's workshops; an electric van; and equipment in the harbor. In addition, the plant produced electricity to illuminate the downtown railroad station and the harbor, as well as to pump water for the steam locomotives.[28]

The East African Railroad Company was a profit-driven private company, and it assumed a central role in the exploitation of the German colony. When it received from the German authorities the first series of requests for electric light, company managers approved the requests. Among the East African Railroad's first customers were various governmental institutions and establishments with substantial state participation: the German East Africa Bank, the Imperial Post Office, the Imperial District Office, the Kaiserhof Hotel, and the German East Africa Company. Approximately two years after acquiring its initial customers, the Railroad Company had attracted an additional thirty-two private customers—and boasted an electricity network that was five kilometers long. At roughly the same time, the company commenced negotiations with the local authorities concerning "the introduction of electric light in the streets."[29] The negotiations went smoothly, and by the end of 1908, the company was able to implement the first, rudimentary street-lighting network serving downtown Dar es Salaam—"financed by the town."[30]

Why did the Railroad Company venture into the business of electric lighting? Available sources offer us a clear-cut (purported) explanation: "This solution is profitable, as it enables us to utilize our electricity plant

[28] *Deutsches Kolonialblatt* 17, 1906, p. 397.

[29] "Dritter Geschäftsbericht der Ostafrikanischen Eisenbahngesellschaft," (1906), p. 3; in Bundesarchiv Berlin Lichterfelde (German National Archives) (hereafter BArch), File No. R 907/11049, "Ostafrikanische Eisenbahngesellschaft Bd. 1 1904–1927." Mr. Markus Schertler was kind enough to provide me with the annual business reports of the East African Railroad Company.

[30] "Vierter Geschäftsbericht der Ostafrikanischen Eisenbahngesellschaft," (1907), p. 5; in BArch, File No. R 907/11049, "Ostafrikanische Eisenbahngesellschaft Bd. 1 1904–1927."

better and have it run more evenly."[31] From a business and engineering point of view, it made sense for the company to complement the need for mechanical power with electric light—both for streets and larger buildings. By combining the production of mechanical power for the company's workshops and the harbor during the day with the provision of electric light at night, the electricity-producing steam engines could be utilized more efficiently, thus increasing the return on investment. From an engineering point of view, this meant that the technicians managed to optimize the steam-engines' so-called load factor.

In his historical analyses, Hughes argues convincingly that the optimization of the load factor was one of the main forces behind the diversification of the power sector in the United States and Europe.[32] When the demand for electric light increased, power companies in many cities began to construct electrically driven tramlines to optimize the use of their electricity-producing plants. Given that customers, primarily, used electric light in the evenings, and the use of trams peaked in the daytime, power companies were thus able to run their electricity-producing plants almost around the clock. In Dar es Salaam, the principle was the same, except the process was largely reversed: first mechanical power, then lighting.

What other comparisons may be drawn between the Global North and the Global South in the case of early electrification? In several studies, historian of technology David E. Nye has shown that in the United States, customers did not necessarily adopt electricity for (primarily) economically rational reasons.[33] Compared to candles, oil lamps, and gas light, electric light was viewed as modern. Similarly, compared to steam engines, electric motors were considered clean. Retailers installed electric lighting to showcase their products, and members of the upper class used electric lighting to signal their wealth. City authorities in North America and Europe made major investments in street illumination to impress visitors and attract wealthy new residents. In the German city of Darmstadt, the Grand Duke

[31] "Vierter Geschäftsbericht der Ostafrikanischen Eisenbahngesellschaft," (1907), p. 5; in BArch, File No. R 907/11049, "Ostafrikanische Eisenbahngesellschaft Bd. 1 1904–1927."

[32] Hughes, "Evolution," pp. 71–73.

[33] Nye, David E., *Electrifying America: Social Meanings of a New Technology, 1880–1940*. Cambridge, MA: MIT Press, 1990. Nye, David E., *Narratives and Spaces: Technology and the Construction of American Culture*. Exeter: University of Exeter Press, 1998.

Louis IV in 1888 decided to have 3,000 lightbulbs installed in his theater—to impress guests and to lend his duchy a certain prestige.[34]

In colonial Dar es Salaam, members of the privileged class also adopted electric lighting to showcase their wealth and demonstrate their connection to the modern world. In 1908, one of the civil servants to the Governor of German East Africa, Baron Albrecht von Rechenberg, complained that the governor's palace still lacked electric light—at a time when "almost all of Daressalam [*sic*]" was electrified. Although this statement was an obvious exaggeration, the civil servant found it "self-evident" that the governor's palace be graced with electric light. After all, this "representative building" was the symbol *par excellence* of German power over its "protectorate." The fact that the palace was situated one-and-a-half kilometers from the center of town could no longer be accepted as a valid argument against hooking it up to the grid.[35]

Given that the electrical network had begun to expand, the East African Railroad Company saw it as its duty to meet the Governor's request. The final offer from the company's Electricity Department, dated November 13, 1908, contains the suggestion that the palace use electric lighting to distinguish itself from other buildings. For the "very cheap" sum of 5,260 rupees, the Electricity Department promised to install 113 lightbulbs and 15 electrical sockets "in accordance with the quality standards of the Society of German Electric Engineers." As a presumptive act of good will, the company provided the cabling free of charge. Immediately after the Governor accepted the offer, the Railroad Company's Electricity Department set about ordering the necessary materials. A total of thirteen chandeliers were to be distributed among the reception rooms, the dining hall, and the veranda. In total, the offer included eighteen pendant lights, six floor lamps, twenty-five wall lights, and one piano light. Visitors to the Governor's palace would almost certainly have been impressed—if not intimidated—by the powerful display of light.

Baron von Rechenberg and the Grand Duke of Hesse-Darmstadt were not the only members of the upper classes who were determined to incorporate electric lighting as a signifier of wealth, if not power. In 1886,

[34] Schott, Dieter, *Die Vernetzung der Stadt: Kommunale Energiepolitik, öffentlicher Nahverkehr und die 'Produktion' der modernen Stadt. Darmstadt—Mannheim—Mainz 1880–1918*. Darmstadt: Wissenschaftliche Buchgesellschaft, 1997.

[35] The quotations in this and the following paragraph are taken from TNA, File No. G7/8: "Verwaltung der Gouvernementsgebäude. Feuerversicherung, 1907–1914."

Sultan Barghash bin Said of Zanzibar ordered the installation of electric light in his House of Wonders (Beit-al-Ajaib).[36] This notably early adoption of electric light showcased the modernity of the Sultan's palace. To impress the Sultan's subjects and guests, the House of Wonders incorporated several modern technologies, an elevator and cast-iron pillars among them. As Nye observes in the case of the United States, electric lighting also appears to have had a particular cachet among the monied in East Africa.[37] A German-language Dar es Salaam newspaper in 1908 testifies to the visual effects cast by the electric lighting: "In the night [everyone] gathered at Bismarckplatz, which was illuminated to complete daylight by arc lamps. Among the rich, green vegetation surroundings of the square emerged the illuminated initials of the [empress's] names."[38]

During the last years of German colonial rule, additional streets were served by electric light, and the number of private customers increased. For the German authorities, electric streetlights became a means of demonstrating the colony's modern status. Household access provided the expat population with the latest comforts—such as the opportunity to iron their clothes with a clean and simple electrical device, rather than a cumbersome coal iron.[39] Emulating the lifestyle of their superiors, low- and mid-ranking soldiers and civil servants also demanded electric light.[40] Both European industrialists and Indian businesspeople profited from the expansion of the electricity network. In contrast, African inhabitants benefitted only when they happened to be employed by the Electricity Department of the Railroad Company. Several factors contributed to the exclusion of the Black population from the network. First, low wages made it impossible for African workers to access electricity in their homes. Second, the living quarters of African people were situated on the outskirts of town, usually beyond the city limits. In fact, the government regarded African people as interim workers only—not as permanent, full-fledged citizens.

[36] Prestholdt, Jeremy, *Domesticating the World: African Consumerism and the Genealogies of Globalization*. Berkeley: The University of California Press, 2008, pp. 109–110.

[37] Nye, David E., *American Technological Sublime*. Cambridge, MA: MIT Press, 1994.

[38] *Deutsch-Ostafrikanische Zeitung*, No. 82, 1908, p. 3.

[39] Straeten, *Capital Grids*.

[40] Hege, Patrick, and Jonas van der Straeten, "Enclaves of Light and Citadels of Profit: Translating Electric Networks for Colonial Dar es Salaam, 1904–1910," *unpublished manuscript*, 2018.

Between 1909 and 1913, electricity consumption in Dar es Salaam more than doubled. The soaring demand forced the Railroad Company to increase power production. In 1913, the company decided to invest heavily in a new building, two steam engines, and the necessary generators.[41] However, "the war interrupted the construction of the new powerhouse for the electricity plant. The two portable steam engines were seized by the enemy shortly before arriving at their destination."[42] As the First World War gained traction, the steady expansion of the electricity network came to an abrupt halt.

Power Disputes: The Electricity Department Versus Customers

At the end of the First World War, the defeated German Reich was required to withdraw from the colonies it had created in Africa. British forces had taken over Dar es Salaam in 1916; three years later, the United Kingdom made the city the capital of Tanganyika Territory. To rebuild the damage to infrastructure and government buildings, the governor in 1920 established the (aforementioned) Public Works Department, a subdivision of which was tasked with running the power system. This division was referred to as both the Electric Light Works and the Electricity Department.

In addition to confronting the repair work on the damage caused by war, the Electricity Department faced problems with what they regarded as unruly customers. On the evening of July 28, 1920, one of its engineers, A.W. Grant, "noticed that the whole shop and exterior lighting of B. Choitram, in the Bazaar, was switched on." This excessive use of electricity raised Grant's suspicions:

> In view of the fact that meters have been tampered with recently, I entered and asked to see their meter. This was not working, and on inspecting it further, I found that the Armature locking device had been released. (The meter is a Siemens Schuckerte)

[41] "Zehnter Geschäftsbericht der Ostafrikanischen Eisenbahngesellschaft," (1913), p. 1; in BArch, File No. R 907/11049, "Ostafrikanische Eisenbahngesellschaft Bd. 1 1904–1927."

[42] "Elfter Geschäftsbericht der Ostafrikanischen Eisenbahngesellschaft," (1914–1919), p. 1; in BArch, File No. R 907/11049, "Ostafrikanische Eisenbahngesellschaft Bd. 1 1904–1927."

> To do this the seal had been removed from the Plug at bottom of [*sic*] meter, and a new seal had been suspended, on a piece of string, from the screw of terminal Box cover.

On leaving the shop of the Bheroomal Choitram trading company, Grant asked the manager to report to the head of the Electric Light Works the next morning, "for inquiry." The customer followed the order, but to no avail. The very same day, the Electricity Department sent an engineer to not only reseal the meter but to cut off Choitram's shop from the power grid.[43]

Bheroomal Choitram was no small undertaking.[44] The main office of this Indian-owned business was situated in Bombay (today's Mumbai), and in addition to its Dar es Salaam location, it owned subsidiaries in Nairobi, Mombasa, and Zanzibar. The Dar es Salaam staff of the Indian trading company made it clear that they would not tolerate what they considered to be unfair treatment on part of the Electricity Department; to force the service provider to reconnect their shop to the electricity grid, Mr. P.P.B. Choitram and one of his colleagues, Mr. V. Bulchand, wrote a three-page letter to the District Political Officer. Like Mr. Patel of the Hindu Public Dispensary, Choitram and Bulchand hoped to find support for their case in the political arena.

Choitram and Bulchand's main argument was that the Electricity Department had merely alleged—not proven—that someone had tampered with the meter's seal. The Choitram company employees suggested that, on the contrary, there was "very strong evidence that the Meter could not have been tampered with by anyone in our employ." And now that the meter had already (ostensibly) been resealed, the turn of events could no longer be verified. What most angered the businessmen was the Electricity Department's move to disconnect the shop from the grid—without allowing them to present their side of the story. Choitram and Bulchand found this treatment unacceptable and even outright unjust: "we can if necessary show him [the Electricity Department engineer] Meters in European offices which have not got the seal which he says our Manager removed or had removed." Unfortunately, there is no way for us

[43] The information and quotes in this and the following three paragraphs are taken from TNA, File No. AB356: "Electric Current – Alleged Theft of."

[44] The spelling of the company's and owners' names varies; sometimes Choithram is used, sometimes Choitram.

to verify whether this statement is true. If so, it stands as an example of the unequal nature of service provision in the colony. Indeed, although many Dar es Salaam dwellers of Indian descent had access to Electricity Department services, Indian people were treated differently from European inhabitants.

The conflict between the Bheroomal Choitram company and the Public Works Department was not a one-off incident. The chaos caused by the war had given local technicians and ordinary users the opportunity to take matters into their own hands. In a written note, a representative of the Electricity Department complains that, "there are a great number of people especially since the war who have sufficient technical knowledge to tamper with meters." Members of the Electricity Department appear to have been particularly suspicious of non-European customers. For example, in July 1920, the Electricity Department accused the Indian-owned Alexander Cinema Company of having manipulated the meter in its theater. One of the department's inspectors, Mr. Mohamed Jumo, had on several occasions observed that the power meter was not functioning. Although Jumo reported this to his employer, no technician showed up to repair the meter. Instead, "the wires were cut by the Electricity Department," making it impossible for the cinema to conduct business.

After being cut off from the grid, the cinema's owners contacted a lawyer. Like Choitram and Bulchand, attorney A.A. Willis decided to approach higher-ranking authorities. In a letter to the District Political Officer, Willis complained of the Electricity Department's discriminatory practices: "A European of course would have insisted on seeing the Head of the Department but I understand that this course is not encouraged at the office of the Electricity Department with non-Europeans." Willis even went so far as to say that the Electricity Department's manager behaved in a way that was "autocratic and high headed."

While the Electricity Department was a "public" enterprise, in that it was a division within the British administration, access to its services was not evenly distributed among the public. Both the hospital anecdote and cinema story illustrate this. For both Bheroomal Choitram and the Alexander Cinema Company, the saga had a positive ending. One police inspector concluded that, "no evidence can be obtained as to when, and who tampered with [the] meters." Given this, the Electricity Department had no prospect of winning a court case against either the Choitram or the Alexander company; both of them prevailed.

As these two cases show, the Electricity Department was in a rather precarious legal position. Unsurprisingly, later in 1920 the newly installed British government issued a "proclamation" to "protect Electrical Energy and Works." Mimicking existing decrees from British India, this legal text stipulated that anyone who:

> (a) disconnects any meter, indicator or apparatus from any electric supply-line without first being duly authorised in that behalf; or
> (b) maliciously injures any meter, indicator or apparatus or wilfully or fraudulently alters the index of any such meter, indicator or apparatus, or prevents any such meter, indicator or apparatus from duly registering; or
> (c) improperly uses energy:
>
> shall be punishable with [*sic*] fine which may extend to three hundred Rupees.[45]

Rules or no rules, the struggle over the provision of electric power would continue.

Electricity Provision: To Privatize or Keep Public

The First World War gave rise to many problems for the inhabitants of Dar es Salaam. Those German citizens who were lucky enough to evade injury found themselves under arrest. British troops forced out many African people from the city. Food shortages propelled others to the countryside.[46] In the early 1920s, however, some people began to return, and the population began to grow again. If we are to believe F.W. Brett, Senior Provincial Commissioner of Dar es Salaam, many "Africans" returned because "they had been accustomed ... [to] the amenities of civilisation." Considering that Brett himself noted that "so called native huts" often lacked basic kitchen facilities, it is unclear what amenities he had in mind.[47]

[45] Proclamation No. 20 or 1920: To protect Electrical Energy and Works, The Tanganyika Territory; in TNA, File No. AB356.

[46] Brennan, James R., and Andrew Burton, "The Emerging Metropolis: A History of Dar es Salaam, circa 1862–2000," in: Brennan, James R., Andrew Burton, and Yusuf Lawi, eds, *Dar es Salaam: Histories from an Emerging African Metropolis.* Dar es Salaam: Mkuki na Nyota, 2007: 13–75, here: p. 29.

[47] "Tanganyika Territory: Provincial Commissioner's Report. Dar es Salaam District, 1921," pp. 8–10, 13; in TNA (without file no.).

One thing is clear, however: Brett did not mean that the African inhabitants of Dar es Salaam had access to electricity.

The growth of the capital became a challenge for the newly founded Public Works Department, given its responsibility for making sure that damaged infrastructure as well as government and other public buildings undergo repair. A 1919 report had outlined the main challenges facing authorities. Many roads were in miserable condition; one passage in the report reads, "street drains are blocked up solidly with a type of sand which sets almost like a cement." The downtown public laundry and the main marketplace buildings were in need of repair, as were various "Mango and Acacia trees." The Chafukoga area flooded repeatedly during the rainy season. Given the pressing need to minimize "mosquito breeding facilities," the Public Works Department set about to improve drainage and fill in swamps.[48]

Apparently, the Public Works Department took seriously its challenge of postwar repair and rebuilding. In his 1922 report, Brett summarized the department's recent achievements:

> The construction of a main drain through a low lying native quarter of the town to take off surface water has been completed. The preparation of the new market is proceeding. The lighting of streets in the town has been improved since the new electric power house has been opened and main thoroughfares in the new European quarters are now lighted by electricity.[49]

One year later, the commercial marketplace where traders sold their products was completed: "Electric light was installed, to admit of business being conducted up till eight o'clock in the evening, an innovation of real value both to salesmen and to the public."[50] In other words, the Public Works Department was a "public" department (in the sense that it was under municipal control), and the Commissioner of the Eastern Provinces obviously meant that it was the Public Works Department's task to provide the "public" (in the sense of the general population) with useful services—including electric light at "public" (generally accessible) places. As

[48] All quotes in this paragraph are taken from "Summary of Report on New Works required in Daressalaam", 30 September 1919; in TNA, File No. 450/39/10.

[49] "Tanganyika Territory: Provincial Commissioner's Report. Dar es Salaam District, 1922," p. 10; in TNA (without file no.).

[50] "Tanganyika Territory: Provincial Commissioner's Report. Dar es Salaam District, 1923," p. 4; in TNA (without file no.).

Brett's remark about "European quarters" indicates, however, this did not mean that electric streetlights were evenly distributed throughout the town.

Economic factors played a central role in expanding the electricity network. In fact, throughout the interwar period, the Colonial Office in Britain usually expressed an unwillingness to finance the maintenance and expansion of power networks overseas.[51] Indeed, the political leadership hoped private capital would take the lead. Expecting to be able to attract private investors, Tanganyika's governor, David Cameron, in 1928 proposed privatization of the publicly owned electricity departments in the territory. The outcome was the 1931 creation of the private companies DARESCO and TANESCO, mentioned above. In my interpretation, these profit-driven companies were "service providers," not "public utilities." Although Cameron claimed that "the public are entitled to expect more efficient and cheap services than they are receiving now," Mchome concludes that Cameron's use of the concept of "public" was limited to the European and Indian segments of the population.[52] African inhabitants were not considered to be part of the "public"—and, given the Tanganyika Territory's implicitly racist practices, Black African people were not provided with "public utilities."

The most important reason for Cameron to privatize electricity provision was likely the sad state of the power-supply system. The following letter excerpt indicates this; E. Dennis, Chief Electricity Engineer of the Tanganyika Railways (once again responsible for the power network), wrote to the Director of Public Works in 1930:

> I would like to call your attention to the general question of Street Lighting in Dar-es-Salaam.
>
> 268 of the Existing Street Lamps have been in use since before 1907.
>
> The bulk of the Standards are so far corroded as to be unsafe.
>
> The cables are at the end of their useful life and it may be thus assumed that the whole of the original Street Lighting Scheme requires re-constructing.
>
> It is becoming yearly more expensive to maintain by means of repairs and in the main, as the system has now become grossly in-efficient for its pur-

[51] Straeten, *Capital Grids;* Mchome, *"Blackout Blues."*

[52] Mchome, *"Blackout Blues,"* p. 34; Mchome here quotes a letter by Sir Donald Cameron to L.M.S. Amery, a Member of Parliament, from Feb. 29, 1928, in: NAK, File CO 691/98/1.

> pose I beg to suggest that you keep in view the necessity of providing funds for its complete renewal.[53]

Dennis argued that the time was right for an overhaul, and indeed, the Railroad Company had imminent plans to convert the entire power system from direct current (DC) to alternating current (AC). These plans emerged after the consulting company Sparks & Partners submitted a report in 1928 on the power system and its mismanagement. On their arrival in Dar es Salaam, the consultants discovered that the manager of the Electricity Department had "only just returned after an absence of some nine months, and the Accountant of the Department [had] gone on leave." Sparks concluded "that considerable expenditure is now necessary to modernise the Undertaking and bring it into satisfactory condition." Investigations showed that the DC system was "not suitable," that various electricity-producing steam-engines were "obsolete," and that the distribution network required repair and maintenance.[54] Other British engineers supported Sparks' conclusions, citing the positive experiences of experts in West Africa, who had implemented the three-phase AC system. The engineers also recommended a particular type of diesel engine, "in considerable use in many Colonies," that provides "satisfaction."[55]

Dennis' letter, and the ensuing correspondence, indicate that actors on the local level in the colonies viewed electricity provision differently from their counterparts in London. Whereas the Colonial Office was unwilling to commit fresh capital to upgrading the power grid, the Executive Officer of the Dar es Salaam Township Authority's Health Office, Dr. R.R. Scott, argued that electricity provision by now had become an "essential public service" in need of government support. "Under pressure from members of the public," the Township understood that it had to act. For instance, a functioning system of streetlighting was "essential for the safety of traffic." The Dar es Salaam police authorities likely had in mind "the present

[53] Letter from E. Dennis, March 14, 1930; in TNA, File No. 18878, "Street Lighting Dar es Salaam."

[54] NAK, File No. CO 691/98/3, "Development of Electric Power: Report by Sparks and Partners." I am grateful to Jonas van der Straeten for giving me access to scanned documents from this file.

[55] Letter by Preece, Cardew & Rider, Consulting Engineers, to the Crown Agents for the Colonies, June 6, 1928; in NAK, File No. CO 691/98/1, "Development of Electric Power." I am grateful to Jonas van der Straeten for giving me access to scanned documents from this file.

epidemic of burglaries" when it asked the Electricity Department to keep "all street lighting in the town ... on all night, instead of half being extinguished at 10 p.m."

Only heavy investment would guarantee the continuous operation and further expansion of the power grid. From March to November 1930, the estimated costs had increased by fifty percent. It is no wonder, then, that the governor put his faith in private enterprise—despite his preference for carrying out negotiations behind closed doors. In a confidential note written toward the end of 1930, and addressed to the Executive Officer of the Township Authority, the Chief Secretary to the Governor regretted "that no definite information in the matter can be given to the general public."[56]

Predictably, the privatization of the power-supply and distribution system did not contribute to more equal treatment of customers. The great majority of African inhabitants lacked the financial means to pay the costs associated with signing a contract with DARESCO. For its part, the company chose to provide electricity only to those customers likely to pay their bills. To avoid breakdowns, outages, and power shedding, the regulatory authorities also forced DARESCO to consolidate and improve the quality of the power system before expanding the grid to new parts of the city.[57]

In fact, high prices seem to have plagued most consumers; even the governor was compelled to try to curb his use of electricity. Upset by the governor's exorbitant outlays for electricity, the government treasurer, Mr. H. Latreille, in 1933 asked DARESCO to "investigate the conditions of consumption of electricity at Government House." After analyzing the building's wiring, lighting system, and electrical appliances, the engineer who carried out the inspection made several recommendations. Given that the electric iron used as much power as the refrigerator, the engineer suggested that residents and guests "discontinue the use of the electric iron and have all ironing done by the charcoal irons." He also "observed that certain lights in corridors and balconies are very darkly coloured and consequently give little light for the current used." A further problem was that the local technicians often replaced 40-watt bulbs with higher wattage bulbs—which automatically increased the governor's power consumption. By far the largest culprit, however, was the electric bathwater heater. This gadget consumed such an enormous amount of energy that the engineer

[56] Information and quotes in this and the preceding paragraph are taken from TNA, File No. 18878, "Street Lighting Dar es Salaam."

[57] Cf. Mchome, *"Blackout Blues."*

recommended "the introduction of a fuel-fired boiler system." For this purpose, the use of wood or coal—rather than the use of electricity—would prove considerably more cost-effective.[58]

Regarding the safety of the electric wiring at the Government House, there was little to criticize. The prospect of safely wiring other centrally located buildings may have been more precarious, however. Electrical engineers argued repeatedly that it would be unsafe to electrify the mud-and-wattle buildings where the majority of the indigenous population lived.[59] The Electricity Ordinance of 1931 and the Electricity Rules of 1932 formalized the division between those who lived in "permanent" and "non-permanent" (also known as "temporary") structures. The rules stated, for example, that safety fuses were to be fitted into a "fireproof" and "moisture-proof" breaker box. Given that African people tended to cook their food over an open fire, and members of the local Zaramo ethnic group primarily used palm leaves for roofing on their houses, the fire hazard in African parts of town was comparatively high. The rules stipulated that all electrical installations had to "conform to the regulations for the Electrical Equipment of Buildings of the Institute of Electrical Engineers of Great Britain"; given this, the engineers' reasoning seems logical. The operative question is: Was it necessary to apply the same rules and regulations in Tanganyika as in Britain, or did these rules represent an instrument of racial segregation?[60] The outcome, in any case, was that African people in Dar es Salaam remained excluded from the electricity grid.

Colonies and the Myth of "Development"

Fire and safety concerns were not the only arguments against the electrification of indigenous people's dwellings; pricing regulations were just as important an obstacle. In fact, on several occasions DARESCO declared its willingness to electrify the African parts of town—but only if the government agreed to support the move financially. The problem was that ordinances and rules prevented DARESCO from offering low-income residents more favorable electricity rates. If the company were to provide

[58] Information and quotes in this paragraph are taken from TNA, File No. 19327, "Electric Lighting in Govt. House."

[59] Cf. Hege and Straeten, "Enclaves."

[60] Both quotations in this paragraph are taken from NAK, File No. CO 691/120/4, "Electricity Ordinance 1931 and Rules 1932." I am grateful to Jonas van der Straeten for having given me access to scanned copies of this file.

African people with electricity at lower prices, European and Indian customers would be entitled to receive the same rates. Still, DARESCO's director, Mr. A.J. Don Small, in 1937 tried to convince the Dar es Salaam authorities to subsidize African customers. Citing several cities in the United Kingdom that subsidized electricity for certain segments of the low-income population, Small contested that no legal objections could be raised concerning such a solution in Tanganyika. In principle, there was no difference between inhabitants of "Council" houses in a British city like Norwich and African dwellers in "Urban Housing Schemes" in Dar es Salaam. Why would it be impossible to "consider a special rate for supply to natives," when low-income families in Britain were getting better deals "on grounds of their inability to pay the higher rate"?[61]

Despite Small's attempt to convince the authorities to provide African dwellers with discount electricity rates, not much happened. The colonial government and the Dar es Salaam Township continued to drag their feet until early 1944, when a letter from the Colonial Office in London forced Governor Sir Wilfrid E.F. Jackson to respond. Having "received an enquiry from a conservative Member of Parliament as to the extent to which provision will be made for the installation of electric lighting in urban housing schemes for Africans in the East African Territories," the Office wanted to know the state of affairs in Tanganyika. The MP in question was Major Abraham Lyons, who in December 1943 had written to Oliver Stanley at the Colonial Office "to suggest that insistence should be made on the necessity for installing electric light in all Urban housing schemes for Africans."[62] Jackson immediately instructed his Chief Secretary to find out if electric lighting had recently been included in such housing schemes. Most provinces countered with the claim that they lacked the capacity to invest in electricity provision. Although DARESCO was allegedly "carrying out experiments as to the types of lighting suitable for the purpose in view," the Provincial Commissioner in the eastern part of the country did not believe that a private company would ever take on such a risky project; only a state-run department would do so.[63]

[61] Information and quotes in this paragraph are taken from TNA, File. No. 24387, "Improvement to Native Houses."

[62] Letter by Lyons December 14, 1943, to Oliver Stanley; in NAK, File No. CO 822/112/6, "Electrical Development." I am grateful to Jonas van der Straeten for having given me access to scanned copies of this file. Cf. also Straeten, *Capital Grids*.

[63] TNA: File No. 24387, "Improvement to Native Houses."

It may surprise readers that providing African people with electricity would be of interest to a Member of Parliament in London—some 10,000 kilometers from Dar es Salaam. What moved Lyons, a Tory lawyer, to rally to this cause? The answer has more to do with politics than with benevolence. As Joseph Hodge and other historians have shown, toward the end of the 1930s, London began to fear that its power over the colonies could be fading.[64] To prevent uprisings and other forms of social unrest, British politicians realized they needed to try to improve the living conditions of their colonial subjects. To this end, politicians and experts redefined the concept of "development."[65] Almost a century of colonial experience had convinced them that colonies did not "develop" automatically from a "primitive" to a "civilized" stage; "evolution" was not an autonomous process. The Colonial Development and Welfare Acts of 1940 and 1945 represented concerted efforts to provide funds to actively assist the colonies' "development," thus enabling them to increase the "welfare" of their inhabitants.[66] Tanganyika had indeed taken the lead in this process: in 1938, Governor Sir Mark Young had already set up a Central Development Committee to formulate an economic development policy. In contrast to later welfare acts, housing projects played only a scant role in the committee's final report. Whereas "a sewerage and drainage scheme for Dar es Salaam" and other "Public health projects" were mentioned under the heading "Works of Public Utility," electricity was not broached.[67] The committee did not deem electricity fundamental to the territory's development.

Although MP Lyons' interventions led the local authorities to investigate the matter in some detail, the governor refrained from promising immediate improvements. Before making any statement, he waited for the outcome of negotiations between DARESCO and the Township Authority. Indeed, almost half a year passed before the Chief Secretary to the

[64] Hodge, Joseph Morgan, *Triumph of the Expert: Agrarian Doctrines of Development and the Legacies of British Colonialism*. Ohio University Press: Athens, OH, 2007, Ch. 6.

[65] The rest of this paragraph is based on Straeten, *Capital Grids*.

[66] Burton, Andrew, "Townsmen in the Making: Social Engineering and Citizenship in Dar es Salaam, c.1945–1960," *The International Journal of African Historical Studies* 36 (2), 2003: 331–365; here: 334.

[67] Tanganyika Territory, *Report of the Central Development Committee*. Dar es Salaam: The Government Printer, 1940, p. 167; in NAK, File No. CO 691/179/15, "Development of Tanganyika: Report of Central Development Committee." I am grateful to Jonas van der Straeten for having given me access to scanned copies of this file.

Governor reassured the Secretary of State in London that the Tanganyika government took the issue seriously: "Matter is under examination with local electric supply company. Action to be taken will depend largely on financial considerations relative to size and location of housing schemes details of which have not yet been settled."[68] In a letter to the Township Authority, DARESCO confirmed that the "Company most certainly share the desire of Members of Parliament in Great Britain to extend the benefits of electricity of the African." Apparently, while legal doubts no longer endured, financial problems remained. When, in August 1944, plans were made to provide "34 African quarters [houses] near the Government School" with electricity, this was nothing more than a drop in the bucket.[69] At the time, approximately 40,000 African people lived in the town.[70]

If nothing else, Lyons' enquiry sparked investigations. Despite the previously raised safety concerns, the authorities were able to merely tick the box showing that "experiments" had been initiated "to put three 40-watt bulbs in a cluster in certain selected existing native houses, in order to see whether this will give the requisite illumination." It was the Township and its administration that pursued the investigations. Taking the lead was electrical engineer A.W. Grant, who had participated in the Choitram case, and who in the 1940s had been employed by the town's Labour Department. According to Grant, there could be no objection to providing "native huts" with one or more fittings: "the services would be simple, no meters would be provided." Customers were expected to pay a flat rate of 2.5 Shillings per installed bulb, amounting to 4 percent of the monthly salary of an average African laborer. Grant explained why such installations would contribute to stabilizing the Empire:

> Let the licensee [DARESCO] make application, without argument, for a supply to hereditary and indigenous Tanganyika Africans, with due provision to exclude claims of other races locally born, at a fair and remunerative rate, which the average Africans can afford, and Government in its role of uplifter of the African cannot but give its enthusiastic support in principle.

[68] Telegram from Jackson to the Secretary of State, July 1, 1944; in NAK, File No. CO 822/112/6, "Electrical Development." I am grateful to Jonas van der Straeten for having given me access to scanned copies of this file.

[69] The quotes in this paragraph are taken from TNA, File. No. 24387, "Improvement to Native Houses."

[70] Brennan and Burton, "Metropolis," p. 38.

In the late nineteenth and early twentieth centuries, Sultan Barghash bin Said and Baron Albrecht von Rechenberg had harnessed electricity to demonstrate power and distinction, endeavoring to impress members of the indigenous population with flagrant displays of light.[71] In the 1940s, by contrast, electric light had become a technology and a service that could—according to colonial reasoning—be used to turn "the African" into a "civilized" citizen, as well as to improve the African person's welfare. The London MP and the Dar es Salaam electrical engineer were united in their belief: technology had political power. According to their mindset, providing non-European people with selected services via technological means would slowly but surely turn said non-European people into satisfied and supportive colonial subjects.[72]

Access to Electricity as an "Amenity"

Until independence, in 1961—and even beyond—only a minority of the African population in Dar es Salaam could afford to pay for electricity for their homes. Despite somewhat more inclusive politics on the part of the British administration in the late 1940s and 1950s, huge socioeconomic differences—and inequities— remained. The fact that African people became recognized as citizens did not guarantee they would have the same degree of political influence and access to infrastructure. For the lowest-income inhabitants, the words of a British colonial administrator from 1932 continued to hold true: "They have no hope of getting electric light." And, in 1954, a consultant to the government concluded: "In Dar es Salaam, the capital town of the territory, the main illumination of the town center [*sic*] is provided from shop window lights, and in most African residential areas, there is no street lighting at all."[73] Almost none of the so-called African Urban Housing schemes introduced after the war included electricity for domestic use—though the schemes did include other amenities, such as piped water and sewerage.

The African people who had permanent jobs and belonged to a small but burgeoning middle class held out hope for electrification; a DARESCO

[71] Cf. the analysis of the demonstrative functions of electrification in colonial Northern Nigeria: Larkin, Brian, *Signal and Noise: Media, Infrastructure, and Urban Culture in Nigeria*. Durham, NC, & London: Duke University Press, 2008.

[72] The quotes in this paragraph are taken from TNA, File. No. 24387, "Improvement to Native Houses."

[73] Both quotes in this paragraph are taken from Ghanadan, "Public Service," pp. 50–51.

representative in 1949 referred to these members of the community as "quite stoic" about waiting for access to the grid.[74] While not all African people in Dar es Salaam likely expected to be able to use electric lights at home in the near future, several individuals did complain to the local administration that they found the streetlighting in their residential areas "inadequate." Sources from the end of the war indicate that, in the minds of African and European citizens alike, electric lighting had started to become one of the expected "amenities and social services."[75] Electric lighting was no longer a luxury of the privileged, a technology with which the European and Indian minorities could demonstrate their power and wealth. An increasingly large proportion of the population had begun to view electricity as a useful service that made everyday life safer and easier.

Still, discourse and practice diverged. Whereas several key actors saw electric light as a necessity, their opinion seldom translated into concrete action. As long as support from London was restricted to one or two Members of Parliament, the Colonial Office was unlikely to channel more funds into expanding the power grid. The politics of "development and welfare" was largely oriented toward industrializing the colonies rather than providing colonial subjects with amenities. Whereas government buildings and some public institutions received electricity, domestic use remained low on the priority list. Only the tiny fraction of African people who belonged to "the nucleus of a middle class" (as one contemporary observer called it) may have benefitted from "a relatively comfortable standard of living"—including even refrigerators and electric fans.[76] Although such "modern amenities" were indeed rare in the African community, they became political instruments that contributed to the integration of an African elite into the colonial system of governance.[77]

The country's postwar development policy gave rise to substantial investments in several hydroelectric dam projects—investment projects too vast for private service providers to finance on their own. In 1960, when conducting a public-relations campaign to raise support for the Hale

[74] *Tanganyika Standard*, 18 June 1949; quoted from Straeten, *Capital Grids*.

[75] Quotes are taken from "Tanganyika Territory: Annual Report of the Provincial Commissioners, 1945," p. 35; in TNA (without file no.).

[76] Molohan, M.J.B., *Detribalization: A Study of the Areas of Tanganyika where Detribalized Persons are Living*. Dar es Salaam: Government Press, 1959, pp. 49–50, quoted from Brennan and Burton, "Emerging Metropolis," p. 44. Cf. Mchome, "'Blackout Blues,'" p. 57.

[77] Leslie, John A.K., *A Survey of Dar es Salaam*. London: Oxford University Press, 1963, p. 151, quoted from Brennan and Burton, "Emerging Metropolis," p. 45.

Falls power plant on the Pangani River, the colonial government emphasized the importance of the project for Tanganyikan industry. Similarly, when independent Tanganyika's (after 1964: Tanzania) first president, Julius Nyerere, inaugurated the dam five years later, he highlighted its role both as a technology that delivered electricity to Dar es Salaam—and as a sign of modernity that united the young nation.[78] The fact that TANESCO (and its subsidiary, DARESCO) had been nationalized a few years after independence underscores the political nature of the project. Hydroelectric dams and power lines had become a means of strengthening the government's legitimacy: electricity remained a politicized technology.

[78] See Straeten, *Capital Grids*; Mchome, *"Blackout Blues,"* p. 78.

CHAPTER 6

Creating "Creole" Cuisine in Latin America: Home Cooks Reinvent *Batánes*

Preparing the Meat

- Kill the heifer, let off the blood, and remove the viscera. Wipe it well with a dry piece of cloth. Cut off the legs at the joints.
- Divide the animal in three parts: shoulders, hindquarters, and ribs.
- Break the carcass open in order that it splits in the middle.

Preparing the Fire

- Make a pit in the ground: 50 centimeters deep, 1.20 m long, and 0.50 m wide.
- Load the hole with enough wood, light the fire and keep it burning for 3 hours in order for the hole to get very hot. [...]
- After three hours, remove the whole fire from the pit and clean it well.

Preparing the Barbeque

- Place the pieces of the calf in the pit, with the hide down—that is, in direct contact with the earth.
- Cover the pieces with a thick [metal] sheet.
- Place the firewood that had been removed from the pit on top of the sheet.
- Keep the fire burning for two hours, adding new wood if necessary.

M. Hård, *Microhistories of Technology*, Palgrave Studies in the History of Science and Technology,
https://doi.org/10.1007/978-3-031-22813-1_6

- During these two hours of cooking, the meat may not be uncovered.
- After the two hours have passed, remove the sheet and take out the roasted meat; place it on a table.[1]

This rustic recipe is taken from an Argentine cookbook, *Especialidades de la cocina criolla* ("Specialties of Creole Cooking"), published in 1958. It describes a marvelously original way of making "traditionally Argentine" *asado* (barbeque): the dish, called *asado con cuero* (barbeque with the hide), does not require the cook to remove the entire hide before cooking. The technologies involved are extraordinarily simple: no grill is required, whether a commercial charcoal grill, a self-made brick grill, or even a rack. The only tools needed are a spade, a metal sheet, and perhaps some matches. A knife for butchering and carving the meat also comes in handy.[2]

It is highly questionable whether most readers of *Especialidades de la cocina criolla* chose to make this *asado* dish. Compiling recipes from the magazine *Mucho gusto* ("Good Taste"), the cookbook catered to a predominantly wealthy urban audience in Argentina. Rather than digging a huge pit in their garden, members of Buenos Aires's upper classes would likely have preferred to use their outdoor grill made of brick. Thus, the cookbook's authors may well have included this country version of "barbeque with hide" for ideological rather than practical reasons. By reproducing such an obviously rural beef recipe, the authors mobilized the image of the gaucho—the brave, unruly South American horseman—to bolster a traditional image of the Argentine nation. What could better symbolize Argentina's cowboy culture than the act of grilling a "two-year-old calf"—hide (*cuero*) and all—in a hole in the ground?

In the nineteenth century, attempts were made to romanticize cowboy life on the South American prairie. However, it was only in the 1940s and 1950s that the exaltation of the gaucho as an icon of authentic Argentine culture reached a peak.[3] Poets hailed the gaucho as a representative of

[1] *Especialidades de la cocina criolla. Seleccionadas por el personal técnica de la revista Mucho Gusto.* 2nd ed. Buenos Aires: Fabril, 1961 (orig. 1958), p. 159.

[2] Alejandra Osorio Tarazona kindly provided me with most of the primary sources and the literature discussed in this chapter. The chapter has profited greatly from her critical comments and corrections.

[3] Adamovsky, Ezequiel, "Criollismo, experiencia popular y política: El gaucho como emblema subversivo," *Anuario del Instituto de Historia Argentina* 18 (1), 2018: e067; Pilcher, Jeffrey M., "Eating à la Criolla: Global and Local Foods in Argentina, Cuba, and Mexico," *IdeAs: Idées d'Amériques* 3, 2012: https://journals.openedition.org/ideas/406

"our freedom" and as a symbol of "nobility and manliness."[4] Gaucho culture was interpreted as distinctly masculine, and the making of *asado* was seen as a typically manly activity. For many in Buenos Aires, barbequing beef became an act of strengthening ties to the vast Argentine countryside. For men in the capital city, making *asado* on an outdoor grill was a way of boosting their identity as hardworking men. By the time "Specialties of Creole Cooking" was published, *asado con cuero* had already assumed the ideological status of *tradicional argentino*—a fact cited by the book's authors.

We find the same heroic gaucho culture in Uruguay. My own former father-in-law, Walter, who grew up in Uruguay, loved to tell stories about his life as a young cowboy on the Pampas. Uruguay's connection to cattle ranching can be traced over several centuries. Accounts of the Uruguayan countryside by foreign visitors feature fascinating descriptions of elaborate meat-based feasts.[5] As the country urbanized, however, the majority of inhabitants in Montevideo faced the same problem as those in Buenos Aires: where to barbeque? With Montevideo's upper-middle class in mind, the magazine *Almanaque* in 1949 explains to those "who feel the cult of tradition" how to build their own grill "corner" in the backyard. The article includes a detailed plan and proposes a "constructive solution for an open-fire barbeque, table, and a mud house or an arbor," all designed according to "the laws of the Creole."[6]

In some regions of Argentina and Uruguay, *asado a la cruz* is an alternative to the labor-intensive, hands-on procedure described in "Specialties of Creole Cooking." *Cruz* means cross, and here, the four limbs of the animal are spread out and hooked onto a metal frame comprised of one long rod and two short crossbars. This double-barred cross is placed either over or adjacent to wood embers or a wood fire. The cook then has the option of either hammering the spit into the ground and angling it toward the fire, or placing the spit horizontally, directly above the embers, propped on two large stones or bricks.[7]

E. Rodrigez Long and Jewel B. Groves, authors of the 1963 cookbook *El asado criollo—Roast Spit Barbeque,* describe this traditional cooking

[4] Quoted from a short poem by A. Don Alberto Vacarezza in the booklet: Aran, Artemio, *¡Pampa!* Buenos Aires: Colección gaucha, 1943, p. 27.

[5] Laborde, Gustavo, *El asado: Origen, historia, ritual.* Montevideo: Ediciones de la Banda Oriental, 2010, Ch. 2.

[6] Quoted from ibid., p. 94.

[7] *Especialidades*, p. 300.

technique in nostalgic terms. Noting that the "spit-roaster used by the primitive River Plate settlers is the same used at present," and that this kind of barbeque evokes "a feeling of festivity," Long and Groves try to reproduce a romantic "image" of the past.[8] In fact, the spit-roasting technique has a long tradition; a source from the early nineteenth century describes how a sheep, still in its hide, is "nailed onto a stick and put in front of the flames."[9]

Like "Specialties of Creole Cooking," Long and Groves's book targeted a wealthy urban audience that could afford the luxury of a "weekend cottage" with enough space to set up a grill or a fireplace. Long and Groves also sought to show a foreign readership how to make "the Argentine Spit Roast"; to this end, their work was published as a dual-language book, with text in both Spanish and English (albeit poorly translated): "An invitation to an 'asado' [the barbeque as an event] produces a feeling of sympathetic cordiality and has a spirit of agreeable camaraderie."

Seen in the context of the trend toward urbanization in the 1950s and 1960s, both "Specialties of Creole Cooking" and *Roast Spit Barbeque* can be interpreted as attempts to revitalize Argentina's rural image. In *El asado criollo—Roast Spit Barbeque*, Long and Groves note that "in recent times" barbeque has become increasingly popular in Argentina. Families compete to make the best *asado*, and amateur cooks exhibit "much ingenuity and subtle cunning to find ways and means of raising or lowering, rotating and other original mechanical contrivances" to improve their grilling stations. Advising cooks on how to design their grills, the authors recommend a surface of at least 110 × 55 centimeters. To allow the cook to control the temperature beneath the meat, only half of that surface should be covered by a grilling rack. The distance between the rack and the area where the wood or coal is distributed should be fifteen to eighteen centimeters. Cooks that "desire to complicate things" may vary this distance by means of a chain attached to a "cog-wheel"; the standard equipment simply included a "long trident, pincers, tongs and poker," as well as a brush.[10]

[8] Long, E. Rodriguez, and Jewel B. Groves, *El asado criollo—Roast Spit Barbeque.* Buenos Aires: Bell, 1963, p. 16.

[9] Quoted from 1824 by representatives of the pope in Argentina; quoted from Arcondo, Aníbal, *Historia de la alimentación en Argentina: Desde los orígenes hasta 1920.* Córdoba, Argentina: Ferreyra, 2002, p. 185.

[10] Long and Groves, *Asado*, pp. 12, 37, 40, 94.

Contemporaneous popular magazines also testify to the upper-class interest in traditional Argentine *asado* culture; this interest took the form of an outdoor private grill corner. Even *Para ti* ("For You"), an Argentine magazine intended primarily for women, includes a detailed description of how to build your own grill. In the 1960 article, which includes a photo of a young couple in their garden, authors Zulema Ciordia and Alfredo Alvarez suggest that "it is not difficult at all" to design and build a barbeque grill. In addition to bricks and mortar, one needs little more than a ruler, a level, and a trowel. Despite the eight drawings that accompany the article, however, I wonder if the average reader had the time and the courage to follow the journalists' instructions to tackle the grill-building project.[11]

The Making of "Creole" Traditions

Historians have shown consistently how social practices turn into traditions. Indeed, some customs have a long history. For example, the use of bread for Holy Communion is no doubt a centuries-old practice in Christianity. Other so-called traditions are much more recent, however. The use of potatoes, originally American, did not spread to Europe until the nineteenth century. Today, potatoes are essential to many allegedly traditional European dishes. For example, Janssons Frestelse ("Jansson's Temptation")—a dish with potatoes, onions, anchovies, and cream—is a staple of the Swedish smorgasbord.

The term "traditional" may also be deployed for political purposes—both in national and colonial contexts. For example, political historians Eric Hobsbawm and Terence Ranger cite cases in which old traditions are mobilized or new traditions created for the political purpose of conferring importance on a particular social practice. Many practices accepted as traditions were coaxed into existence with a specific message in mind; this phenomenon is summarized elegantly in the title of a well-known book edited by Hobsbawm and Ranger: *The Invention of Tradition*.[12] For example, in his contribution to this volume, Bernhard S. Cohn shows how the British Crown in the nineteenth century tried to legitimize its power in India by staging ceremonial events intended to position Britain as a direct

[11] *Para ti* (Buenos Aires), Vol. 39, Issue 1998 (25 October 1960), pp. 41–43.

[12] Hobsbawm, Eric, and Terence Ranger, eds, *The Invention of Tradition*. Cambridge: Cambridge University Press, 1983.

heir of former Mughal emperors and Indian princes.[13] By reproducing age-old rituals, the British appropriated established traditions to achieve their own ends.

Accordingly, is there an empirical case to be made for classifying Argentine *asado* as a tradition "invented" to further political aims? I believe so. After all, the authors of "Specialties of Creole Cooking" classified *asado con cuero* as a rustic, "traditionally Argentine dish" associated with gaucho culture. In doing so, the authors engaged in an apparently conscious effort to foster the image of Argentina as a "cowboy" culture.[14] And when Long and Groves claimed that a "roast on the hide is … a native symbol of … Argentine customs," they ignored the contributions of Argentina's indigenous people to the country's history.[15]

The authors of *Especialidades de la cocina criolla* appear to define the "traditional" recipe in contradictory ways. On the one hand, the authors write, "many of the dishes we now consider 'traditional' mirror the influence from distinctly colonial cuisine and large migratory movements"—from Europe, Asia, and Africa. On the other hand, the authors point out that Latin American cuisine is characterized by domestic ingredients such as cassava, corn, potatoes, sweet potatoes, and an array of chili peppers. Thus, implicit in the authors' definition of "traditional" cooking is the idea of combining global elements with regional or local ones.[16]

A clarification: the term *criollo/criolla* in this context differs from what some readers may associate with the concept of "Creole" cuisine. The authors of the Latin American cookbooks cited in this chapter refer to *la cocina criolla*—the creole kitchen—rather than to specific Louisiana-style dishes like gumbo and jambalaya, also known as "Creole." The authors' concept was much broader, perhaps even diffuse. The explicit purpose of *Especialidades de la cocina criolla*, for example, was to document "authentic recipes […] from all over Latin America." Although the authors do not clearly define "creole," they often use the concept to describe recipes and dishes they deem either typically Latin American or typical of a nation or region. Some of these creole dishes do indeed seem uniquely Latin American, and they lack English translations: "'tamales', 'humitas',

[13] Cohn, Benhard S., "Representing Authority in Victorian India," in: Hobsbawm and Ranger, *Invention*, pp. 165–209.

[14] *Especialidades*, pp. 9–11.

[15] Long and Groves, *Asado*, p. 98.

[16] This paragraph is based on *Especialidades*, pp. 9–11.

'hallacas', 'anticuchos', 'asado con cuero', 'chipas.'" The cookbook even includes a recipe for "Aztec soup," which, according to the authors, has "been passed on from generation to generation without changes." Most of the recipes, however, are acknowledged to be "derivations" of foods from other parts of the world; the recipes are for dishes with origins in distant places that have been modified in the Americas. Reflecting a multitude of influences, these dishes are the outcome of complex acculturation processes.[17]

Sociologists of technology have analyzed extensively the ways in which consumers and users incorporate new implements and appliances into their daily lives, making technologies "their own."[18] For example, in the 1970s, when the microwave oven was first introduced in Europe, family members appropriated this device to suit established cooking and eating habits, and they "assimilated" the oven into their daily routines.[19] Similarly, the makers of Latin American cuisine have throughout the centuries exhibited astounding flexibility, appropriating elements of cuisines from the other sides of the Atlantic and the Pacific—while continuing to use indigenous ingredients and to employ familiar cooking techniques.

Originally, *criollo/criolla* ("creole") referred to a man or woman of Iberian ancestry born in the Americas. Politically and socially, the term was employed to distinguish American-born people with a European family tree from indigenous people, Africans, and first-generation immigrants from the southwest of Europe. As such, the term creole referred to an identity that was established predominantly—though not exclusively—on racial grounds. Over time, however, the meaning of "creole" changed considerably, eventually interpreted differently in various parts of Latin America. In some contexts, *criollo* carried positive connotations; in others, it evoked negative associations.

When used in the context of cooking, *criollo* has come to refer to a specific attribute: the adoption of ingredients and techniques from elsewhere into Latin American cuisine. Food historian Rebekah E. Pite describes accurately the complex usage of the term *criollo* as a combination

[17] This paragraph is based on ibid., pp. 9, 11, 46.

[18] Lie, Merete, and Knut Sørensen, eds, *Making Technology Our Own? Domesticating Technology into Everyday Life*. Oslo: Scandinavian University Press, 1996.

[19] Johansson, Birgitta, *Ny teknik och gamla vanor: En studie av mikrovågsugnens introduktion*. Linköping: Linköping University, 1988, p. 139.

of "localism, ethno-regional difference, and melting-pot-style fusion."[20] In some examples of creole cooking, the domestication process has advanced to the point of rendering invisible the "foreign" influence on the dish. In these cases, creole food has become more or less synonymous with local, regional, or national dishes. Historian Jeffrey M. Pilcher points to the ideological and social elements of creolization, writing that "national cuisines emerged throughout Latin America not from the rejection of the global in favor of the local but rather through a blending of the two in a culinary sensibility that combined patriotism and cosmopolitanism in pursuit of social distinction."[21] In nineteenth-century Cuba, for example, creole cuisine came to denote European dishes that had been modified to include indigenous American root crops such as potatoes and cassava—as well as African ingredients like okra and yam. A Cuban cookbook from the mid-nineteenth century offers a "creole beef tongue" recipe, which could have been taken from any number of European cookbooks—except that this version of the dish features plantains, the type of banana used for cooking.[22]

When cookbook authors in Latin America mobilized the concept of *criollo*, giving certain recipes and ingredients a positive connotation, those authors invented culinary traditions. The reference to creole traditions made it easier for them to successfully introduce new ingredients and cooking technologies. For example, although original recipes for *asado* specified that it be cooked over an open fire, more recent recipes were modified to fit modern living conditions and modern technologies—without sacrificing the reference to the *asado*'s creole heritage.[23] Although the modern homemaker might not have access to an outdoor barbeque area, it was still acceptable to make a traditional *asado* in a gas or electric oven. As we will see later in this chapter, the familiar "fusion" character of cross-cultural culinary practices went hand in hand with a "fusion" approach to kitchen technology.

[20] Pite, Rebekah E., "La cocina criolla: A History of Food and Race in Twentieth-century Argentina," in: Alberto, Paulina L., and Eduardo Elena, eds, *Rethinking Race in Modern Argentina*. New York: Cambridge University Press, 2016: 99–125, here: 102.

[21] Pilcher, "Eating," p. 1.

[22] Ibid., p. 9.

[23] Concerning the use of the concept of "creole" in the history of technology, see Edgerton, David, "Creole Technologies and Global Histories: Rethinking How Things Travel in Space and Time," *HoST: History of Science and Technology* 1, 2007: 75–112, here: 102.

Eclectic Cooking

Another well-known cookbook, published in 1890 in Argentina, testifies to the multicultural character of Latin American cuisine: *Cocina ecléctica*. To create *Cocina ecléctica*, journalist and public intellectual Juana Manuela Gorriti solicited recipes from people throughout Latin America and beyond. A truly collective effort, *Cocina ecléctica* includes recipes contributed by nearly two hundred women—and one man. The majority of contributors came from Argentina and Peru; others hailed from Bolivia, Uruguay, and Mexico. Several contributors were from as far away as Spain, France, Ireland, and the United States.

The fact that *Cocina ecléctica* was produced almost exclusively by women made it a pioneering endeavor.[24] The vast majority of the contributors belonged to the upper classes (Gorriti had at one time been married to the then President of Bolivia, Manuel Isidoro Belzú). The recipes reflect the contributors' elite socioeconomic status, as well as their geographical diversity. In some of the recipes, authors refer to the "foreign" origins of their dishes. In other cases, authors refer to the deeper cultural roots of the recipes they present. For example, one contributor from Peru acknowledges the "Incan" origins of the well-known dish called *Pachamanca*, which is meat—anything from chicken to vicuña to guinea pig—baked with hot stones. Still other authors refer to their dishes with nationalistic pride: one contributor, from La Paz, claims that "nowhere else" in the world is rabbit prepared better than in Bolivia. By presenting a recipe for a "Breton tortilla," yet another contributor shows that traditionally Latin American dishes can be given a European touch.[25]

References to servants seldom appear in "Eclectic Cooking"; yet it was indeed the servants for upper-class families who did most of the actual cooking, employing the necessary skills and techniques.[26] Accordingly, the book's contributors pay minimal attention to the technologies used in the cooking process. There are notable exceptions, however. One outstanding

[24] Women's early contributions to the culinary literature of Argentina are analyzed by Caldo, Paula, *Mujeres cocineras: Hacia una historia sociocultural de la cocina Argentina a fines de siglo XIX y primera mitad del XX*. Rosario, Argentina: Prohistoria, 2009.

[25] Gorriti, Juana Manuela, *Cocina ecléctica* (edited by Camilla Cattarulla). Buenos Aires: La Crujia, 2014 (orig. 1890), pp. 109, 177, 207.

[26] Austin, Elisabeth, "Reading and Writing Juana Manuela Gorriti's *Cocina ecléctica*: Modeling Multiplicity in Nineteenth-Century Domestic Narrative," *Arizona Journal of Hispanic Cultural Studies* 12, 2008: 31–44.

example is an elaborate recipe for *humitas* (also spelled *humintas*)—a variety of corn wrap with pre-Hispanic roots—submitted by Gorriti's daughter, Edelmira Belzú de Cordoba. Like her father, Edelmira's husband, Jorge Córdoba, was President of Bolivia, the latter for two years. Edelmira likely maintained a kitchen staff of servants, in which case she would have taken on the role of manager instead of cook. Indeed, her instructions for how to make *humitas* read as though written by a keen observer rather than a hands-on cook. Unlike her fellow contributors, however, Edelmira credits the "Indian" cooks appropriately, and she describes in detail the techniques and technologies involved in the cooking process.[27] A common first step in preparing *humitas* is to grind corn. According to Edelmira, this is done manually, either "on a batán or, if there is none at hand, in a stone mortar." A *batán* has two main parts: a large, flat millstone, and a heavy, rounded hand-stone; the *batán* enables the user to grind all manner of foodstuffs. Together with the three-stone hearth—used for traditional cooking by most people around the world—the *batán* played a pivotal role in pre-Columbian and colonial cooking culture.[28] Grinding corn, coffee, herbs, and spices on a *batán* was a central part of women's work.

Edelmira's recipe calls for combining the ground corn with salt, sugar, pork fat, and chili peppers. Three spoonsful of this mixture are then placed on two crisscrossed corn leaves, and the leaves are folded and bound with twine made from the agave plant. The bundles are then cooked "in a pot, in the oven, or in a *guatia*." Of these three methods, Edelmira strongly recommends the *guatia* (also spelled *huatia*), an Incan stone-and-earth oven. Edelmira describes in detail the *guatia*'s design:

> The Indians place one [stone] on top of another on the ground and make a small, ten-centimeter-deep hole at the top. Here, they light a fire and stoke it vigorously to heat the stones.
>
> When the fire has reached the appropriate temperature, those taking part in the process wrap their hands in strips of burlap. With skill and speed, they dismantle the oven.

After removing the upper stones, the cooks insert the *humitas* in the hole, then reassemble the oven, thus "forming a wall of heated stones" around

27 Gorriti, *Cocina*, pp. 81–84.

28 Fernández-Souza, "Grinding and Cooking: An Approach to Mayan Culinary Technology," in: Ayora-Diaz, Steffan Igor, ed., *Cooking Technology: Transformations in Culinary Practice in Mexico and Latin America*. London: Bloomsbury, 2016: 15–27.

the *humitas*. After approximately an hour, the corn bundles are ready—and the cooks referred to as "Indians" once again remove the stones with "the same skill and speed."[29] Unfortunately, all the "Indians" remain anonymous, and Edelmira romanticizes their work.

Edelmira's recipe for *humitas* references a far more complex cooking method than the one described in *Especialidades*, published seventy years later. Reflecting the urbanization process and the emergence of modern apartment buildings with indoor kitchens, the 1958 version of the recipe specifies merely boiling the corn wraps in a pot with "abundant water" and coarse salt.[30]

National or *Criollo* Desserts

Gorriti's concept of eclectic cuisine squares well with Pite's broad definition of *cocina criolla*—creole cuisine. As we saw in Edelmira's contribution, Gorriti's book depicts how Latin Americans domesticated foodstuffs from elsewhere, *and* it provides detailed descriptions of indigenous cooking methods and techniques. This is a remarkable achievement for an author writing in 1890. To my mind, Gorriti and her circle of contributors represent pioneers of what is now known as "fusion" cooking.

In most parts of the world, we are surrounded by evidence of fusion cuisine's popularity; we can observe pronounced global influences in a multiplicity of forms, from cookbooks featuring inspired recipes to restaurant and food-bar menus offering creatively cross-cultural dishes. That said, I maintain that we should not exaggerate the "melting-pot" character of either *cocina criolla* or the contemporary culinary phenomenon called fusion cooking. Instead, we can recognize that the many culinary technologies did not exactly hybridize or "fuse"; rather, culinary technologies have existed side by side, and these technologies have been used in an eclectic—rather than a uniform or orthodox—way.

Consider an example from Peru: for roughly two decades, Peruvian cuisine has been circulating globally, at an ever-increasing pace. *Ceviche* (marinated raw fish or seafood) and *pisco* (a white brandy made from grapes) have been at the forefront of this trend. Various dishes made with corn, quinoa, and potatoes have also been popular. Peruvian restaurants have cropped up in small and large cities; diners have learned to enjoy

[29] Gorriti, *Cocina*, pp. 83–84.
[30] *Especialidades*, p. 117.

pachamanca and *tamales* (sweet corn wraps filled with fresh corn kernels rather than ground corn).

Interestingly, the recent proliferation of Peruvian food outlets does not appear to hinge on their serving "authentic" Peruvian food. On the contrary, most contemporary restaurants that feature so-called Peruvian food are noted for the creative ways in which their chefs combine different cooking styles and technologies. Cookbook authors often cite the disparate roots of Peruvian cuisine, referring to the mix of Incan, Spanish, and African traditions.[31] In fact, Peruvian cuisine incorporates the foodways of many more regions and cultures. For example, in the nineteenth century, French cuisine was a decisive influence on Peruvian cooking—especially among the upper classes. In the postcolonial period, waves of migration from Italy, China, and Japan brought foodways from those countries. For example, Italian immigrants introduced to Peru new vegetables such as chard, spinach, and broccoli. In the second half of the nineteenth century, Chinese immigrants opened their first tiny eateries, known as *chifas*. These establishments served *lomo saltado* (stir-fried beef) and *arroz chaufa* (fried rice).[32] After the turn of the twentieth century, the Japanese *cocina nikkei* proved especially compatible with other culinary traditions—to such an extent that one food historian even called it "miscegenation gastronomy" (*mestizaje gastronómico*).[33] The use of soy sauce with sautéed or grilled beef is an iconic example of this kind of cuisine, as is the substitution of rice wine with *pisco* in some recipes.

Arguably, the international ascent of Peruvian cuisine dates to the 1960s. In 1965, for example, *Cocina peruana* ("Peruvian Cuisine") was published.[34] The book's subtitle reveals its modest scope within the Peruvian kitchen: *Cocina China, dulces criollos, secretos de cocina*—"Chinese Cuisine, Creole Desserts, and Cooking Secrets." Indeed, the book reveals the secrets of Peruvian cuisine to be its Chinese and other

[31] Cuadra, Morena, and Morena Escardó, *The Big Peruvian Cookbook: 100 Delicious Traditional Recipes from Peru*. New York: Skyhores Publishing, 2019.

[32] Balbi, Mariella, ed., *Los Chifas en el Perú: Historia y Recetas / Chinese restaurants, 'Chifas,' in Peru: History and Recipes*. Lima: Universidad San Martin de Porres, 1999.

[33] Guardia, Sara Beatriz, "Gastronomía peruana del siglo XX: Historia e identidad," in: Villavicencio, Maritza, ed., *Seminario historia de la cocina peruana*. Lima: Universidad de San Martin de Porres, 2007: 229–242, here: 237. Cf. also Cuadra, Morena, and Morena Escardó, *The Peruvian Kitchen: Traditions, Ingredients, Tastes, and Techniques in 100 Delicious Recipes*. New York: Skyhorse, 2014.

[34] *Cocina peruana: Cocina China, dulces criollos, secretos de cocina*. Lima: Bendezu, 1965.

transcultural elements. The book's anonymous authors devote almost half of it to Chinese cuisine and so-called creole sweets (*dulces criollos*). The other half of the book contains purportedly domestic recipes, such as "Arequipa Stew," "Peruvian Ram's Leg," and "Lima Soup."[35] True to the regional nature of many of the recipes, the authors of *Cocina peruana* call for the reader to add "an ounce of chocolate" to the Arequipa Stew and ground *achiote*—a spice made from the red seed of the annatto tree—to the ram's leg.[36] After all, both cocoa and *achiote* (*bixa orellana*) are indigenous Latin American plants. Further, it is likely impossible to find a European cookbook from the 1960s that specifies chocolate as an ingredient in a meat dish. In the Global North of the 1960s, chocolate was associated almost exclusively with desserts and sweets.

A closer look at *dulces criollos* in "Peruvian Foods" reveals the fusion nature of Peruvian cuisine, encompassing the ingredients used and the names of some of the dishes. *Queque*, for example, is a sponge cake made with margarine, sugar, milk, eggs, flour, food coloring, seeds of the vanilla bean, grated orange peel, and raisins. Although these ingredients approximate those of sponge cake recipes found elsewhere in the Spanish-speaking world, the name of the dessert is unique to the region. In Spain, for example, this kind of cake is known as *bizcochuelo* rather than the Peruvian word, *queque*. In addition to the recipe for *queque*, the authors of *Cocina peruana* offer readers a recipe for *bizcochuelo*. Given that the Peruvian version of *bizcochuelo* does not contain food coloring, milk, orange peel, or raisins, it can be considered a simplified, usually smaller, and probably somewhat drier version of *queque*. The authors also provide a third, related recipe, this one for *cake*. To make *cake*, the authors call for adding more sugar, eggs, and cocoa to the *bizcochuelo* mixture. Compared to baking *bizcochuelo*, baking *cake* was a more elaborate and more expensive process.[37]

Other recipes for "creole desserts" in *Cocina peruana* contrast more sharply with their Spanish counterparts. One example are *alfajores de Moquegua*, round cookies known by the same name in Latin America as in parts of Spain. Despite sharing a name, the cookies are prepared differently according to the region. In Andalusia, for example, *alfajores* do not contain filling, and they are baked in the oven. In contrast, the Latin American version of *alfajores* calls for the cookies to be filled with a sweet

[35] Ibid., pp. 8, 33, 48.

[36] Ibid., p. 8.

[37] This paragraph is based on ibid., pp. 103–105.

paste and fried in butter, rather than baked in the oven. The cookie dough itself is not unusual, as it contains flour, water, sugar, eggs, and baking soda. One of the cookie's suggested fillings is unusual by Spanish standards, however: the Peruvian recipe specifies *camotillo*, a mixture of mashed sweet potato, sugar, and grated orange peel. This addition of *camotillo*—a traditional Peruvian dessert that may also be eaten on its own—distinguishes the Peruvian version of *alfajores* as unique.[38]

The *dulces criollos* section of *Cocina peruana* also includes several recipes for *mazamorra*, a corn custard.[39] This renowned dessert has identifiable indigenous roots in pre-Hispanic times. Closest to the original is the recipe for what the authors call *mazamorra morada*, "purple *mazamorra*." This dessert is made in two steps, first, by boiling a purple corncob—along with pineapple peel, dried cherries, cinnamon, and cloves—for as long as three hours. The mixture is then strained, and the reduced liquid is combined with pieces of pineapple, quince, and peach, as well as sugar, sweet potato flour, and lemon. This is boiled until it reaches the consistency of a custard.[40]

Although they employ the concept "creole cuisine," the authors of *Cocina peruana* do not refer explicitly to *mazamorra morada* as a dessert indigenous to Latin America. However, the authors offer two additional *mazamorra* recipes, both of which have traceable European roots. For these recipes, the authors specify using wheat flour instead of corn flour, as well as cow's milk, a product that did not exist before the first Europeans brought cattle to the Americas. Other ingredients are sugar, anise, and cinnamon; one recipe calls for adding "a good amount of egg yolk." When boiled, the result is a sweet dessert reminiscent of *blancmange* and other thick puddings.[41]

In their recipe for a sweet dish called *frejoles colados* (literally "strained beans," not to be confused with various savory Central American dishes with similar names), the authors of *Cocina peruana*—intentionally or not—do not mention the African roots of the dish. According to some sources, *frejoles colados* arrived in the Americas with the first slave ships.[42]

[38] This paragraph is based on ibid., pp. 106–108.

[39] The "(indigenous) ethnic origins" of *mazamorra* is verified by Pite, "Cocina," p. 107.

[40] This paragraph is based on *Cocina peruana*, p. 119.

[41] The recipes for "Mazamorra de leche" and "Mazamorra de yemas" are found in *Cocina peruana*, pp. 108, 112.

[42] The African heritage of this dish is mentioned in: Hinostroza, Rodolfo, *Primicias de cocina Peruana*. Léon: Everest, 2006, p. 226.

The recipe calls for cooking peeled beans without salt and mashing them with milk. The mixture is then pressed through a sieve; sugar is added, and the resulting paste is boiled. The dish is served garnished with toasted sesame seeds.[43] *Frejoles colados* is still served today, particularly in Lima.

Nor do the authors discuss the historical roots of the recipe for *ranfañote*.[44] Although culinary historians have not reached agreement on the origins of this bread pudding, they do concur that *ranfañote* was invented during colonial times. Consisting of slices of fried bread mixed with sugar and honey, *ranfañote* reflects the collective sweet tooth that developed as sugarcane cultivation became more widespread in Latin America. Like people in India (see Chap. 4), people in Peru traditionally preferred a semi-soft form of unrefined sugar, rather than white, refined sugar. In Latin America, these brownish sugar cakes are known as *panela*, *chancaca*, or *piloncillo*. The name *panela* has to do with its form: it is usually sold in pieces similar to bread (*pan*) slices. Like Indian *gur* or *jaggery*, this kind of sugar is produced locally, directly from pressed and boiled sugarcane juice without centrifugation or further refinement. In *Cocina peruana*, *chancaca* is a required ingredient of *ranfañote*.

Other ingredients also exemplify the colonial—and even global—origins of *ranfañote*. Specifically, *chancaca* is mixed with honey, cloves, cinnamon and salt; chopped walnuts and coconut flakes are also added. Interestingly, both clove and cinnamon are originally Asian plants that arrived in the Americas during the colonial era. This may be the case for the coconut tree, as well, although evidence suggests it may have arrived on the west coast of South America from the Pacific region as early as pre-Hispanic times.[45]

As we have seen, many of the "creole desserts" described in *Cocina peruana* were indeed cross-cultural creations. The desserts were colonial in terms of having emerged in the colonial era. In the case of Peru, "colonial" refers to the pre-1820s period, before the country achieved independence. These "creole desserts" may also be considered precolonial or pre-Hispanic, in that they featured ingredients unique to the Americas. Further, these desserts are global insofar as they include ingredients and

[43] This paragraph is based on *Cocina peruana*, p. 110.

[44] Cf. ibid., p. 107.

[45] Baudouin, Luc, and Patricia Lebrun, "Coconut (*Cocos nucifera* L.) DNA Studies Support the Hypothesis of an Ancient Austronesian Migration from Southeast Asia to America," *Genetic Resources and Crop Evolution* 56 (2), 2009: 257–262.

cooking techniques from different parts of the world, from Asia, to Africa, to Europe. *Dulces criollos* denoted desserts with a "global" background but with specific Latin American elements. Creole recipes were the outcome of active appropriation, or what can also be referred to as domestication processes. Defined as such, "creolization" is a practice that strives to turn something external, and perhaps strange, into something recognizable and familiar.

The Modern Kitchen

Traditional cookbooks seldom discuss in detail the tools, implements, and appliances that readers are expected to use. Authors of nineteenth-century cookbooks took it for granted that cooks knew which techniques to employ when preparing a dish. One notable exception is the category of cookbooks written to promote a particular technology—gas stoves and electric stoves, for example. In Germany, *Das elektrische Kochen* ("Electric Cooking") provided readers with recipes as well as explanations of how to become familiar with the modern electric stove and oven. The first few editions of *Das elektrische Kochen*—issued from 1936 onward—targeted as the audience housewives relatively inexperienced with electric gadgets. This successful book is still available under the title *Das blaue Kochbuch* ("The Blue Cookbook")—the fifty-sixth edition—as of this writing.[46] Originally, it was published by the Berlin Electricity Company; today by the electrical engineering association VDE.

Cookbooks designed to address electric-stove owners are also part of Latin America's literary culinary history. In the mid-1930s, Peru's Association of Electric Companies (Empresas Eléctricas Asociadas) published a cookbook—*Recetas selectas para cocinas modernas* ("Selected Recipes for Modern Kitchens")—to address the "electric kitchen." The book claims electric appliances "bring happiness to the home" and save women's "silken and elegant hands" from coming in contact with wood, coal, and "primitive" open fires; working in an electric kitchen is so simple that "even a child can handle" it. Further, the advantage of this form of "modern cooking" can be summarized by the attributes of "comfort,

[46] Petersen, Sonja, "*Das elektrische Kochen* – Die Vollelektrische Küche als Leitbild moderner Haushaltsführung," *Food & History* 11, 2013: 75–106; HEA: Fachgemeinschaft für effiziente Energieanwendung e.V., ed., *Das blaue Kochbuch: Das Koch- und Backbuch für Anfänger und Fortgeschrittene.* 56th ed., Berlin: VDE Verlag, 2020.

beauty, hygiene, efficiency, affordability." A colorful drawing at the beginning of the book highlights these benefits. The illustration features a chic-looking woman wearing elegant evening attire. She is browsing a magazine titled "Modern Kitchens" (*Cocinas modernas*). In the background we see a spacious, antiseptically clean kitchen, with an electric stove and oven, situated beside a sink and several cupboards. The message is unmistakable: by adopting electric appliances, the modern woman can enjoy the luxuries of cleanliness and leisure; modern machinery can free women from the hardships of traditional homemaking.[47]

In the 1930s, the Association of Electric Companies framed electric appliances in a modernist discourse, asserting that people were "living in the century of electricity," claiming electrical power would ultimately bring about "progress"—and even a "revolution." Accordingly, the standard wisdom suggested that a "modern home" was required to be connected to the electricity grid; lacking electric power was likened to riding a mule rather than a streetcar. Further, the authors suggested that "one child and an electric kitchen" were the keys to "the happiness of a modern marriage." Cognizant of Peru's low connection rates at the time, the book's authors chose to project its vision of the modern home into the near future, predicting it would not be long before an "enormous quantity of electrical apparatuses" were in use.[48]

The modern home depicted in "Selected Recipes for Modern Kitchens" looks typically European. In the drawing mentioned above, the elegant woman has red hair, casting her as Northern European, perhaps British. Another illustration features a cook of clearly European descent—with white skin and pink cheeks. Although the authors suggest that an electric stove allows the homemaker to "prepare the best Creole dishes," many of the recipes associate electricity with modernity—and with Europe. For example, the ambitious homemaker is taught to use the modern electric stove to cook "Spanish Fish," "English Roast Beef," and "Italian Tripe."[49]

In Argentina, electric kitchen appliances were marketed in a similar way. In addition to referring to the modernity and Europeanness of such technologies, arguments about sanitation played an important role. Electric and gas stoves also had the advantage of being equipped with an oven.

[47] *Recetas selectas para cocinas modernas*. Lima, Peru: Empresas Eléctricas Asociadas, 1936, pp. 8, 10, 43.

[48] Ibid., pp. 7–9, 53.

[49] Ibid., pp. 10, 30, 36, 42, 56, 58.

Before ovens made their way into urban households, it was customary for Argentine housewives and maids to bring bread and cakes to the nearest bakery to be baked. A food historian lists numerous implements that entered upper- and middle-class homes from the late nineteenth century onward: "machines to grind meat, to make pasta, to grate bread or cheese, to extract juice from meat, to beat eggs."[50] Most of these implements were designed to accommodate European- and North American-style cooking. As in the Global North, urban households in Latin America began to embrace both mechanization and electrification. The process was especially swift in Buenos Aires, where, in 1947, as much as forty percent of all households had a refrigerator.[51]

In the 1950s, *Para ti*, the Argentine women's magazine, ran articles that weighed the advantages of modern kitchen appliances.[52] Advertisements by manufacturers of both electric and gas stoves depicted women homemakers in much the same way as the cookbook published by the Association of Electric Companies. In 1960, *Para ti* included a Longvie Company ad that rhapsodizes about a "moist cloth" that wipes clean the stove's "smooth" surfaces.[53] An accompanying drawing shows an energetic, fashionably dressed woman serving dinner to her family; eagerly awaiting their roast chicken are the woman's husband, teenaged son, baby—and Grandma. In contrast to wood and charcoal stoves, gas stoves promised to liberate women from the hardest household chores—and to promote culinary and aesthetic success.

In an ad that highlights the centrality of *asado* in Argentine culture, the Longvie Company claims its "oven is especially suitable to grill exquisite roasts." Whether your guests want their meat cooked rare or well done, the Longvie oven will "please the most demanding" palates. Arguing that it "knows much about stoves," the manufacturer promises home cooks the most delicate *asados*.

[50] Arcondo, *Historia*, p. 222.

[51] For Argentina in general and Buenos Aires in particular, see Ballent, Anahi, "La 'casa para todos': Grandeza y miseria de la vivienda masiva," in: Devoto, Fernando, and Marta Madero, eds, *Historia de la vida privada en la Argentina: La Argentina entre multitudes y soledades*. Vol. 3: *De los años treinta a la actualidad*. Buenos Aires: Aguilar, Altea, Taurus, Alfaguara, 1999: 19–47.

[52] Pérez, Inés, "Corazón de hojalata, hogar de terciopelo: La cocina, epicentro del mundo doméstico (Mar del Plata – Argentina, 1950–1970)," *Asparkía: Investigació feminista* 21, 2010: 105–128.

[53] *Para ti*, Vol. 39, Issue 1995 (4 October 1960), p. 37.

The Longvie ad crystallizes one of the most debated ideas in the history of household technology: Whom is household technology intended to help? In 1983, Ruth Schwartz Cowan published the foundational, aptly titled book *More Work for Mother.*[54] According to Cowan, in premodern times, men and women in North America had a relatively equal hand in carrying out household tasks. In the nineteenth century, as the processes of industrialization and urbanization advanced, however, household labor came to be defined more exclusively as women's work. Increasingly, women were consigned to the domestic sphere, while men worked outside the home as wage earners. By necessity, women took on household tasks previously performed by husbands and older sons—chopping firewood and hauling water, for example.

So-called modern household equipment was marketed as convenient for the homemaker. Cowan documents how manufacturers of coal, gas, and electric stoves tried to convince women of their products' efficiency. Manufacturers reasoned that merely turning a knob was far simpler than splitting firewood, for example. Cowan refutes manufacturers' claims of efficiency, however. According to Cowan, in adopting allegedly time-saving appliances, "housewives" met with higher expectations from others.[55] No longer satisfied with a mere stew for dinner, husbands now demanded three-course meals, served with condiments and side dishes. Home economists also pressured women to change their cooking practices and adopt the latest insights from nutrition science. Psychologists added to some women's To-Do lists by admonishing mothers to spend more time with their children. The outcome, at least in the United States: *More Work for Mother.*

To my mind, Cowan's thesis speaks directly to the 1950s Longvie ad—and to the sales campaigns for household appliances in 1950s Argentina.[56] Consider the details of the Longvie ad and the magazine in which it appeared. *Para ti* catered to an urban audience. The majority of *Para ti* readers came from Buenos Aires and belonged to the middle class; most readers likely lived in apartments without easy access to a "grill corner." Even if they had wanted to, husbands would hardly have been able to pull

[54] Cowan, Ruth Schwartz, *More Work for Mother: The Ironies of Household Technology from the Open Hearth to the Microwave.* New York: Basic Books, 1983.

[55] Ibid., p. 64.

[56] Pérez, "Corazón," p. 121, touches upon the debate about time-saving household devices in Argentina.

off an elaborate, "traditionally Argentine" barbeque on a regular basis. So, why not delegate to their wives the responsibility for the *asado*? While meat roasted in a gas oven is not identical to meat barbequed on an outdoor grill, the Longvie ad assured home cooks that their diners would nonetheless pronounce their food "exquisite." And given that women were said to "know a great deal about cooking," their families—as well as their dinner guests—would be pleased with the results of gas-oven cooking. Whereas the men of the house had been responsible for the weekend *asado*, it was now the woman of the house who was charged with delivering the goods at mealtime. Intentionally or not, Argentine mothers were expected to do more work.

"Modern" recipes and cooking techniques, adapted to urban ways of life, began to appear as early as the interwar years. For example, it was in this period that renowned Argentine chef Antonio Ganzaga proposed a "modern" version of *asado con cuero*.[57] His recipe is considerably more practical than the far-fetched one cited at the start of this chapter. Ganzaga relieves the cook of having to use the whole animal—and of digging a hole in the ground. Instead, Ganzaga suggests using a standard grill (*parrilla*). Another Argentine culinary luminary of the era, cookbook author Mercedes Cullen de Aldao, provides another modern alternative to making *asado*: roasting it in the oven.[58] And in 1940, the upscale magazine *Vosotras* ("Yours") explained to its (mostly women) readers how to set up a simple barbeque on weekend camping trips and picnics.[59] In the following decade, the authors of *Especialidades de la cocina criolla* reworked several recipes in which meat or fish is cooked in the oven—presumably baked, roasted, or broiled—rather than grilled outdoors. Clearly, the revised recipes were an attempt to accommodate the cooking styles of modern urban life.

The trend toward increased expectations and higher quality standards was also apparent in the interwar period. Emphasizing the scientific and technological aspects of homemaking, Josefina Brusco in 1934 defined the kitchen as the true "laboratory of the home," a place where the housewife is responsible for providing her family with the "necessary nutrition." Brusco was a teacher at the so-called Official Institute of Domestic Science,

[57] Pite, "Cocina."

[58] Marta, *La cocinera criolla y recetario curativo doméstico*. 18th ed. Barcelona: Luis Gili, 1928, p. 24.

[59] *Vosotras* (Buenos Aires) 6 (223), Jan. 5, 1940, p. 83.

in Lima. In a comprehensive book titled *El amigo del hogar* ("The Friend of the Home"), Brusco discusses in detail what modern homemakers needed to know—and which technologies they needed to apply. *El amigo del hogar* is more than a collection of recipes; it is a treatise on how to manage a modern home.[60] "The Friend of the Home" joined the ranks of its counterparts in the United States and Europe—most notably Christine Frederick's *Household Engineering*, published in 1920.[61] In choosing a scientific approach, Brusco managed to differentiate her book from the traditional household manual "Kitchen Lessons," published in Lima more than a decade before.[62]

Brusco begins her book by describing how kitchens should be designed and maintained. Good ventilation is a must, and "order and cleanliness are indispensable." Wooden cupboards and sets of drawers are recommended; zinc and marble may be used for table surfaces. The list of items "one must have in the kitchen" extends to several pages and exemplifies the complexity of modern cooking and housekeeping. In addition to more than twenty different kinds of pots and pans, Brusco asks homemakers to acquire a "machine to grind meat, almonds, etc.," various cake-pans, knives of different sizes, a sieve, a potato ricer, scales, spatulas, beaters, cutting boards, measuring spoons, a mortar-and-pestle, scissors, a corkscrew, a can opener, a water kettle, a coffee maker, and more.[63] Only two of the items listed are authentically Latin American artifacts: a traditional *chocolatera* (hot-chocolate maker) and a *batán* (the pre-Columbian millstone mentioned above).

In the process of advancing "Domestic Science," Brusco positioned the modern kitchen as a scientific laboratory, and the homemaker as a technician. Brusco emphasizes the science of cooking by explaining the nutritional value of various foods—and the health benefits of specific vitamins and minerals—over the course of eight pages. According to Brusco and her cohort, the modern homemaker was expected to be a knowledgeable, versatile professional. Brusco's book, as well as the institute she created,

[60] Brusco, Josefina, *El amigo del hogar: Cocina criolla y extranjera con recetario para enfermos y convalescientes.* Lima: T. Scheuch, 1934, p. 7.

[61] Frederick, Christine, *Household Engineering: Scientific Management in the Home.* Chicago, IL: American School of Home Economics, 1920.

[62] *Lecciones de cocina: Contienen las más deliciosas y fáciles recetas de cocina como también consejos prácticos para el hogar.* Lima: Imprenta del estado, 2nd ed., 1921. The anonymous author, "a woman from Lima," was probably Zoila M. Bernales.

[63] Brusco, *Amigo*, pp. 7–10.

reflect a transnational trend in the interwar period toward transforming homemaking into a modern occupation, and the housewife into a scientifically educated and technologically trained expert on the home.[64]

Promoting Modern Cooking

Most present-day historians of technology argue that women—whether as homemakers, teachers, promoters, or architects—actively co-designed the modern kitchen.[65] Women's associations discussed the place of modern technology in the home, and they often acted as mediators between manufacturers and individual consumers. Members of the newly minted profession of Home Economist took on mediating roles as expert instructors as well as authors of cookbooks and other manuals.[66] This was the case in many countries of the Global North as well as in selected Latin America countries, including Peru and Argentina.

In April of 1947, the Racing Club of Buenos Aires—a well-known football club—organized a special event completely unrelated to sports: the "Ladies' Subcommittee" of the club offered a cooking class. Admission was free, and the event was more of a demonstration than a class. The idea was to show the audience how to use electric stoves. To this end, the committee had invited representatives of the Argentine Electricity Company (Compañía Argentina de Electricidad). The event was a success, and later the same year, Luisa Tosi de Rojo, Secretary of the Ladies' Subcommittee, asked the State General Gas Administration (Dirección General del Gas del Estado) if it would be willing to host a similar "demonstration." Convinced by the advantages of gas as a "fast, economic, and safe" form of energy, Rojo was "eager to collaborate" with the Gas Administration.[67]

[64] Landström, Catharina, "National Strategies: The Gendered Appropriation of Household Technology," in: Hård, Mikael, and Andrew Jamison, eds, *The Intellectual Appropriation of Technology: Discourses on Modernity, 1900–1939*. Cambridge, MA: MIT Press, 1998: 163–188.

[65] Oldenziel, Ruth, and Karin Zachmann, eds, *Cold War Kitchen: Americanization, Technology, and European Users*. Cambridge, MA: MIT Press, 2009.

[66] Chatriot, Alain, Marie-Emmanuelle Chessel, and Matthew Hilton, eds, *The Expert Consumer: Associations and Professionals in Consumer Society*. Aldershot: Ashgate, 2006.

[67] The source material for this and the following paragraph is taken from the General National Archives of Argentina (Archivo General de la Nación): File No. 218 Re, "Dirección General del Gas del Estado," retrieved by Alejandra Osorio Tarazona.

Rojo's request made its way slowly through the Gas Administration bureaucracy. It took two months before she received a reply. Finally, the Gas Administration agreed to organize "demonstrations in the culinary arts" at the Racing Club. "Determined" to show that the use of gas was "very rational and very economical," decision-makers at the Administration chose the Department of Gas Sales to plan and host the so-called cooking class. Members of that department were to collaborate with Señora Petrona C. de Gandulfo, the organization's Director of Rationalization. Gandulfo was responsible for fostering "new culinary uses for gas"; she was also the head of the organization's School of Home Economics.

At the time, Gandulfo was already a high-profile media personality in Argentina. Her food columns were published regularly in popular magazines; she was a frequent presence on radio cooking shows; and she had published several editions of her own cookbook: *El libro de Doña Petrona*. The book, first published in 1934, included close to one thousand recipes. Subsequently, Gandulfo was known simply as Doña Petrona.[68] Before joining the newly founded state-owned Gas Administration in 1946, she had been employed as an *ecónoma* (home economist) by the private Compañía Primitiva de Gas. Together with the magazine *El hogar*, this company organized well-attended public cooking demonstrations in Argentina's larger cities on a regular basis. By 1947, then, Gandulfo could look back on nearly two decades as a writer and instructor. The Ladies' Subcommittee of the Racing Club could not have recruited a better representative.[69] Later, in the 1950s, Gandulfo's media celebrity increased: she became a television star.

Historian Rebekah E. Pite demonstrates how Gandulfo's career and areas of culinary expertise mirror Argentine society from the 1920s to the 1960s. Before the Second World War, for example, Gandulfo's recipes evince a pronounced European style, and her cookbook addresses primarily the upper echelons of society. This readership could afford to buy Gandulfo's pricey book, as well as the costly ingredients required for many of its recipes. Presumably, Gandulfo's readers also had the time to cook elaborate meals, and many of them could afford to employ domestic help.

[68] Gandulfo, Petrona C. de, *El libro de Doña Petrona: Recetas de arte culinario*. Buenos Aires: no publisher, 1934.

[69] Pite, "Cocina." Information about Doña Petrona in this and the following two paragraphs is largely taken from Pite, Rebekah E., *Creating a Common Table in Twentieth-century Argentina: Doña Petrona, Women, and Food*. Chapel Hill: University of North Carolina Press, 2013.

After the Second World War—when Juan Perón ascended to the presidency—Gandulfo shifted her attention to a middle-class audience. Simultaneously, the US influence on her cooking became more pronounced, as she had entered into agreements with US-based food manufacturers. Reflecting the general trend toward *cocina criolla* in the late 1950s and 1960s, Gandulfo now strove to include in her repertoire traditional Latin American ingredients and recipes.[70]

Doña Petrona's demonstrations and publications echoed Argentine modernity in several ways. First, her association with gas providers and manufacturers of gas stoves highlights the importance of advanced technology in modern, urban homes. Gandulfo's cookbook calls for kitchen appliances that require electricity or gas—not just ovens, but also ice-cream makers, for example.[71] Second, Gandulfo professed to address "modern" women specifically.[72] In South America as in Europe, this modern woman was portrayed as poised between being a good homemaker, a conscientious mother, and a professional wage-earner.[73]

Recognizing that middle-class women had neither the time nor the money to prepare elaborate daily meals, Gandulfo in 1962 issued a new cookbook: *Recetas económicas*.[74] Here, she recommended using processed foods to save time and more affordable ingredients to save money; Gandulfo also suggested serving fewer dishes for week-night dinners.[75]

Gandulfo's recipes and television programs represent the era's concept of "modern cooking" in other ways, as well. For example, when recommending the use of processed foods, she often refers to their time-saving potential—simultaneously supporting the modern food processing industry. Canned products made by Swift & Company feature in her TV shows and publications, as did ingredients manufactured by the Royal Baking Powder Company.[76] In other words, Doña Petrona mediated between consumers and the gas industry—as well as between consumers and the world of modern retailing.

[70] Pite, *Creating*, Introduction.

[71] Ibid., p. 55.

[72] Ibid., p. 145.

[73] Cf. Heßler, Martina, *"Mrs. Modern Woman": Zur Sozial- und Kulturgeschichte der Haushaltstechnisierung*. Frankfurt am Main: Campus, 2001.

[74] Gandulfo, Petrona C. de, *Las recetas económicas de Doña Petrona*. Buenos Aires: Distal, 2002 (orig. 1962).

[75] Pite, *Creating*, p. 155.

[76] Cf. ibid., Ch. 2.

Latin America and the Multicultural Kitchen

Unquestionably, some of the appeal of Latin American cuisine stems from its mingling of indigenous traditions and influences from all over the world. Foodways have always accompanied migrants in their journeys across borders and oceans; people of different backgrounds have always learned cooking techniques from each other. To quote Gorriti, Latin American gastronomy has acquired an "eclectic" character.

Latin American cooks and commentators alike refer frequently to this form of "melting-pot" eclecticism in terms of "creole" cuisine. As we have seen, the concept of *criollo* has many different connotations and denotations. Historians and social scientists have used a range of terms and definitions to capture "*criollo*" and "creole." Some scholars use "hybridization" to describe the process by which Latin American cuisine has developed.[77] Others investigate how cooking practices undergo a "mixing" process in which dishes emerge.[78] One cultural historian, writing at the start of the twenty-first century, refers explicitly to a "general tendency [...] toward a mestizo—in some places called a criollo—cuisine."[79] Another historian prefers the term "mosaic" to describe the multifaceted character of Latin American cuisine.[80] To my mind, the important point is the emphasis on the active "incorporation" of multicultural elements—from ingredients to cooking techniques—used simultaneously with established elements. This is a process in which the previously unknown is made familiar. It is also a process in which customary ingredients and techniques do not simply disappear. Rather than rejecting external cultural influences, Latin Americans have often creatively adopted foodstuffs and technologies from elsewhere, to fit their own traditions. This is exactly what Doña Petrona did, in her recipe for *humitas*, when she replaced the customary corn husks with cannelloni, the large, tubular pasta.[81]

Fusion cuisine requires a fusion kitchen, a space where technologies of various genealogies are employed. Latin American anthropologists observe

[77] Cf. Archetti, Eduardo P., "Hibridación, pertenencia y localidad en la construcción de una cocina nacional," *Trabajo y sociedad* 2 (2), 2000, https://www.unse.edu.ar/trabajoysociedad/ARCHETTIFINAL.htm [accessed Aug. 22, 2022].

[78] Laborde, *Asado*, p. 111.

[79] Bauer, Arnold J., *Goods, Power, History: Latin America's Material Culture*. Cambridge: Cambridge University Press, 2001, p. 186.

[80] Arcondo, *Historia*, p. 225.

[81] Pite, *Creating*, p. 73.

that, today, even in the remote countryside, "cooks combine traditional and modern techniques and technologies." *Comizcals*—traditional wood-burning ovens—often coexist with modern ovens.[82] Similarly, in wealthy urban homes on the Yucatán Peninsula, one may find, side by side, "wooden mortars to prepare Caribbean *mofongo*, woks, crystal recipients [bowls] for Russian caviar, pizza stones, *terrine* pots, and Dutch ovens."[83] Upper-class kitchens in this region may also feature both an electric mixer and a *metate*—a large stone of pre-Hispanic origins used for manual milling.

The fusion character of many Latin American kitchens is a long-standing phenomenon. Toward the end of the nineteenth century, Argentine shops displayed an array of kitchen equipment. As with electric coffee machines, most products were of European—especially Italian—origin; after the First World War, imports from the US increased.[84] Still, the Italian influence continued, as evident in a 1940 advertisement in which a Buenos Aires manufacturer launches a "practical machine" for "the making of tagliatelle and ravioli."[85]

The increased use of "modern" kitchen equipment did not erase indigenous dishes and technologies. On the contrary, mechanization processes sometimes supported traditional cuisine. In the 1920s, in the wake of the Mexican Revolution, the government actively promoted the installation of gasoline- and electricity-powered corn mills in the countryside. Adapted from standard European cereal mills to grind hominy (precooked corn grains, called *nixtamal*), the *nixtamal* mill was championed as a means of freeing women from the relentless daily labor of the *metate*. Although rural inhabitants had earlier objected to its introduction, and while it continued to be contested, the *molino de nixtamal* found a substantial market in the interwar period. And after the war, the *molino* was joined by the *tortilladora*, a mechanized appliance for making corn *tortillas*, a staple of indigenous Latin American cuisine.[86] To my mind, it makes sense to call mechanical corn mills and *tortilladoras* "fusion" technologies.

[82] Magaña González, Claudia Rocío, "Technologies and Techniques in Rural Oaxaca's Zapotec Kitchens," in: Ayora-Diaz, *Cooking*, pp. 55–67; here: 55, 64.

[83] Ayora-Diaz, Steffan Igor, "Home Kitchens: Techniques, Technologies, and the Transformation of Culinary Affectivity in Yucatán," in: idem., *Cooking*, pp. 85–98; here: 93.

[84] Arcondo, *Historia*, p. 222.

[85] *Vosotras* (Buenos Aires) 6 (223), Jan. 5, 1940, p. 69.

[86] Bauer, *Goods*, pp. 145, 188–191.

In line with Cowan's scholarship and argumentation, I question whether these implements eased the burden for women homemakers. Indeed, the material I have presented in this chapter supports Cowan's more-work-for-mother thesis in the case of Latin American cooking. This chapter is also meant to reinforce the claim that historians of technology can help us to better understand the creolization of Latin American foodways over the last centuries. The making of *cocina criolla* required the simultaneous adherence to old—as well as the openness to new—manual techniques and material technologies. Creolization meant more than adopting technologies from other regions; it also meant the continuous use of well-established domestic kitchen implements and appliances. Sometimes, traditional recipes were modified to fit modern technologies. At other times, the continuous use of traditional technologies required the modification of imported dishes and recipes.

The concept of "hybrid technology" strikes me as unhelpful to my interpretation of creole cuisine in Latin America. I find Edgerton's concept of "Creole technology" similarly unhelpful. I have not come across examples of cooking technologies which, having been abandoned in the Global North, were further developed in the Global South. It is indeed difficult to find cases in which foreign and domestic technologies have developed into true "hybrids": a *molino de nixtamal* does not include any elements from the *matate*. As food for thought, I suggest that the place where fusion cuisine is being practiced—the kitchen—takes on a fusion character. What we can learn from Latin American social scientists and historians is that the old and new continue to coexist in the kitchen, creating a living mosaic of cuisines.

PART III

Postwar Innovations

CHAPTER 7

Earning a Living in Urban Africa: Maintaining the "Native Beer" Economy

> What was Village II like in 1970? If you had walked down the main street of Village II from the Juja Road, you would have passed along the uneven road between houses whose frontages had been adapted as shops and trading premises; past the carpenters' workshops and an open-air barber's shop under a temporary shelter and more shops whose windows open on to the road, with a shutter hinging inwards and downwards to make a counter. Notices bore messages such as "Hey Girls Welcome" and "*Heshima Maisha Mpya Hoteli ya Ruku*" (Luke's Prestige New Life Hotel), or 'Muturi Co-operative Brothers' Super Investments', advertising bars and shops. Men stood in groups chatting, and others gathered round a mechanic mending a car or repairing a bicycle, or a tailor sitting at his treadle sewing machine.[1]

This account could apply to many large villages or small towns around the world. The businesses and landmarks would have different names, of course, and not all of these enterprises could necessarily be found along the main road. But, in most sizable settlements, you would indeed find carpenters, tailors, and hairdressers, as well as repair shops, bars, and eateries (*Heshima Maisha Mpya Hoteli ya Ruku* was most likely a restaurant rather than a hotel, as stated). Like Village II, the majority of large

[1] Hake, Andrew, *African Metropolis: Nairobi's Self-help City*. London: Sussex University Press, 1977, p. 150.

M. Hård, *Microhistories of Technology*, Palgrave Studies in the History of Science and Technology,
https://doi.org/10.1007/978-3-031-22813-1_7

settlements would have had a centrally placed administration building; a venue for social events; water-provision facilities; and a school.

The description of Village II is taken from a book published in 1977 by Andrew Hake, a British clergyman who had spent more than a decade in Nairobi as a project adviser to the National Christian Council of Kenya. Village II—also called Ndururu, after its charismatic chairman, Mr. Ndururu Kiboro—was in fact defined as a "shanty town" or "squatter area" at the time.[2] Defying standard definitions of "slums," Ndururu was a lively place—probably more so than other settlements of comparable size. Although Ndururu consisted of only 300 houses at the time, its economy was highly diversified. A 1969 survey reported the presence of five bicycle repairers, eight barbers, and twelve shoemakers. In addition, forty-five people working as tailors or seamstresses, and six people offering laundry services. The construction business was represented by ten blacksmiths, twenty masons, and fifteen carpenters. Two people even made a living by repairing watches. Almost ninety women stated that they were actively involved in "home industries," which included needlework, weaving, knitting, and basketmaking. Some of these endeavors were organized cooperatively.

More than four hundred Ndururu dwellers were employed by larger enterprises and institutions in the area: a credit society, a hotel, a women's cooperative, and the Social Hall. The village center encompassed a nursery school and a dispensary, as well as several small shops, and local-administration offices. The village's diversified economy accommodated the majority of the population; only forty people earned their living beyond the boundaries of Ndururu. Some of the village's dwellers worked in a nearby hospital, while others worked on-site as day laborers, watchmen, and domestic servants, for example.[3]

Ndururu's tailoring trade illustrates the versatile nature of the settlement's economy. Garment makers worked in their own homes, in tiny shops—or they rented a couple of square meters in the street itself. In many cases, no distinction was made between homes and shops: "Many so

[2] Archives & Special Collections, School of African and Oriental Studies (SOAS), University of London: Papers of Rev. Andrew Hake [hereafter: SOAS, Hake], PP MS 46: File No. 4:12, "Slums/Squatters/Shanty Towns."

[3] This paragraph is based largely on SOAS, Hake, PP MS 46: File No. 4:17, "Mathare Valley Housing Project," in addition to Etherton, David, ed., *Mathare Valley: A Case Study of Uncontrolled Settlement in Nairobi*. Nairobi: Housing Research and Development Unit, University of Nairobi, 1971.

called shops are obviously living quarters, containing beds and personal effects." Businesses were small. Tailors typically had a single employee or apprentice; records show that one person employed five assistants in a somewhat larger shop. The majority of garment workers had neither received institutionalized training nor taken part in a formal apprentice system. Only one worker each had attended a vocational school, a missionary school, and a course offered by the Singer Sewing Machine Company. Most businesses rented—rather than owned—treadle sewing machines. Several people in the tailoring trade had access to cutting machines, as well as packing, buttonhole, pressing, shirt-folding, and overlock machines. Their work was demand-driven, and the tailors did not keep inventories. Locals brought the fabric of their choice and commissioned the seamstress or tailor to make whatever garment they needed. Alternatively, customers requested mending services. Some garment workers did piecework, as subcontractors to larger companies beyond the village. A survey from the time shows that the garment workers' biggest concern was the competition they faced from "the second hand clothes dealers who keep shops in the village."[4]

The Stereotype of the "Slum"

Ndururu was one of five villages that, together, comprised a region which continues to be known as Mathare Valley, located seven kilometers from downtown Nairobi. From the description of Ndururu at the start of this chapter, even a close reader might infer that, in 1970, the Mathare Valley was well established. It was not. Regarding the Valley as a hideout for Mau Mau rebels, British forces in the early 1950s had cleared the whole area; it took until the 1960s for people to resettle. By 1970, Ndururu could certainly not be characterized as middle class; nor could it be called working class. Given the socioeconomic circumstances of Ndururu, contemporary commentators may well have called the settlement a "shanty town."[5] "Squatter area" was a similarly demeaning term. And still others would have surely used the word "slum."

In analyzing Ndururu's identity as a "shantytown" or "slum," it makes sense to consider other urban spaces with the same designation. In a

[4] The material in this paragraph is based on SOAS, Hake, PP MS 46: File No. 5:110, "Clothing Industry."

[5] *The Sunday Post* (Nairobi), March 9, 1969.

critique of this attribution, historian Alan Mayne points out that, since its first emergence in the nineteenth century, the word slum has been a "stereotype, a fantasy" that commentators apply to perpetuate a negative view of particular settlements. Mayne concludes convincingly that "slum talk misrepresents poor neighbourhoods and their residents as being deficient, disordered and unchanging."[6]

Urban spaces branded as "slums" usually receive bad press. For example, the *favelas* of Rio de Janeiro are *not* known for the diverse, creative ways in which dwellers earn their living. Rather, *favelas* are marked as places of extreme violence and danger, and scholars writing about the *favelas* reproduce the one-sided story line that these settlements are inhabited by "teeming masses trapped in misery," and that they "remain an officially unrecognized and illegal part of the city."[7] Similarly, when analyzing *bustee* dwellers of Dhaka or *geçekondu* inhabitants of Ankara, students tend to highlight their low incomes and cramped living conditions.[8] In Peru, squatter areas are euphemistically called "young villages" (*pueblos jóvenes*), implying makeshift buildings and temporary circumstances.[9]

In all of these cases—and in general—urban spaces deemed "slums" are associated primarily with: violence and social tension; sanitation problems; inadequately built, transient structures; unemployment and poverty; insufficient provision of resources. The standard story line is one of deficiency and vulnerability.

Scholarly and popular works on so-called shantytowns and slums reproduce this narrative. In Mike Davis' widely read *Planet of Slums*, the deficiency story line looms large: "Instead of cities of light soaring toward heaven, much of the twenty-first-century urban world squats in squalor, surrounded by pollution, excrement, and decay." In his characteristically enraged tone of voice, Davis suggests that Nairobi and other "poor

[6] Mayne, Alan, *Slums: The History of a Global Injustice*. London: Reaktion Books, 2017, pp. 9–10.

[7] Pino, Julio César, "Sources on the History of Favelas in Rio de Janeiro," *Latin American Research Review* 32 (3), 1997: 111–122; here: 111.

[8] Mahmud, Shihabuddin, and Umit Duyar-Kienast, "Spontaneous Settlements in Turkey and Bangladesh: Preconditions of Emergence and Environmental Quality of Gecekondu Settlements and Bustees," *Cities* 18 (4), 2001: 271–280.

[9] Bonilla, Francisco A., *Pueblos jóvenes y urbanizaciones populares: Organización vecinal en Lima metropolitana y resto del Perú*. Lima: Ed. Mercurio, 1972.

megacities ... are stinking mountains of shit."[10] Similarly, Christine Bodewes' account of Kibera, another Nairobi "slum," is meant to spotlight the desperate situation of the inhabitants: "Lacking access to sanitation, drains, rubbish collection, clean water and electricity, the living environment is polluted and unsanitary."[11]

Redefining the "Slum"

Is there a counter-narrative to the slum as a putrid urban space? Where can we read about the knowledgeable craftspeople, the canny restaurant owners, the innovative businesspeople who inhabit these areas?

Indeed, some scholars have remarked on the intrinsic creativity of people in Nairobi's "squatter areas." For example, in a 1968 study, geographer Kenneth Gordon McVicar highlighted "the wide range of economic activities frequently present in such areas." In his investigation of daily life in Pumwani—another of Nairobi's low-income areas—McVicar documents the diversity and flexibility of the people and the economy. The inhabitants of Pumwani experienced "considerable latitude in the application of ingenuity to small scale economic activity." On the one hand, there were those who went to work every morning in other parts of the city: wage-earners, pupils, and students. Unskilled people were employed as "waiters," "cleaners, turnboys, or messengers." Among those who had undergone some kind of training were welders, bakers, and truck drivers. Several members of the Luhya people worked in the printing business. On the other hand, there were "also the people who rely on their own devices—those who, driven by ambition, or despair, have hacked out a niche in the highly competitive African economy." During his extensive fieldwork, McVicar came in contact with "bottle collectors and shoe makers, hawkers, traders, and beer sellers, witch doctors, and prostitutes," in addition to musicians, hairdressers, and tailors—most of whom were active in the Pumwani settlement itself.[12]

[10] Davis, Mike, *Planet of Slums*. London: Verso, 2006, pp. 19, 138.

[11] Bodewes, Christine, "Chang'aa Drinking in Kibera Slum: The Harmful Effects of Contemporary Changes in the Production and Consumption of Traditional Spirits," *African Journal of Drug & Alcohol Studies* 9 (1), 2010: 49–55; here: 49; cf. also idem., *Parish Transformation in Urban Slums: Voices of Kibera, Kenya*. Nairobi: Paulines Publications Africa, 2005.

[12] The quotes in this paragraph are from McVicar, Kenneth Gordon, "Twilight of an East African Slum: Pumwani and the Evolution of African Settlement in Nairobi," *PhD Dissertation*, University of California Los Angeles, 1968, pp. x, 195–198, 268.

In an effort to develop more accurate, more objective terminology, urban geographers, sociologists, and planners have reframed so-called slums as "informal settlements." For example, writing about Lagos, Nigeria, Matthew Gandy suggests the city is "shaped to a significant degree by informal and unplanned settlements."[13] While Gandy refrains from using the specific term slum, his characterization nonetheless emphasizes the precarious conditions of the inhabitants, many of whom were unemployed or underemployed, compelling them to squat on land owned by others. The genealogy of the concept of informality goes back to Keith Hart's work on unofficial economic activities in Accra, Ghana.[14] In his 1973 article, Hart investigates and classifies various types of economic activities in what he called "the informal sector."[15] Notably, the attribution "informal" carries a clear advantage for city authorities: this designation relieves municipal authorities of their obligation to provide a given settlement with the necessary services. That means sanctioning the failure to connect low-income settlements to the centralized water-supply network and the electricity grid—services to which only "formal" settlements are entitled.[16]

Other scholars have framed "informality" in terms of "networks."[17] In her introduction to an anthology on African labor, Ilda Lindell asserts that it is impossible to draw a clear line between formal and informal structures.[18] In a paper coauthored with Mats Utas, Lindell suggests that "webs of social relationships" guarantee access to economic resources and help provide political influence. Further, the focus on "social networks" may enable us to surpass the unhelpful distinction between "informality" and "formality."

[13] Gandy, Matthew, "Planning, Anti-planning and the Infrastructure Crisis Facing Metropolitan Lagos," *Urban Studies* 43 (2), 2006: 371–396, here: 386.

[14] Hart, Keith, "Informal Income Opportunities and Urban Employment in Ghana," *The Journal of Modern African Studies* 11 (1), 1973: 61–89.

[15] Ibid., p. 68.

[16] Compare my critical discussion of the concept "public utility" in Chap. 5.

[17] Lindell, Ilda, and Mats Utas, "Networked City Life in Africa: Introduction," *Urban Forum* 23 (4), 2012: 409–414, here: 409.

[18] Lindell, Ilda, "Introduction: The Changing Politics of Informality—Collective Organizing, Alliances and Scales of Engagement," in: idem., ed., *Africa's Informal Workers: Collective Agency, Alliances and Transnational Organizing in Urban Africa*. London: Zed Books, 2010: 1–30.

I agree that the concepts of a "web" or "network" are useful for analyzing the activities in so-called slums and similar urban spaces. But I argue that none of the existing concepts or nomenclature captures satisfactorily the reality and the character of such settlements; a new term is needed.

Ndururu as a Flexible Settlement

The majority of inhabitants in Ndururu were low-income. Some were even destitute. Many of the people who worked in the vocations listed at the top of this chapter had a hard time making ends meet. But these economic facts of life in Ndururu, circa 1970, cannot be conflated with the cluster of conditions associated with "slum."

On the basis of archival material from Nairobi in the decade immediately after Kenya gained independence in 1963, I suggest that the concept of the "flexible settlement" better describes communities like Ndururu. Voluntarily or not, the inhabitants of such settlements must adjust to constantly changing conditions; one day may bring a living wage, and the next day may end in going to bed hungry. The threat of police knocking at the door—or the government clearing the entire site—is ever-present. Although a fairly high degree of turnover in the population creates a certain anonymity, it also means that there is a constant influx of new ideas and initiatives. Despite the precariousness of life in a squatter community, people learn how to cope with the state of temporality. To earn a living in such a place, one must be both flexible and creative. One settlement-dweller in Addis Ababa phrased it succinctly: "We are good at surviving."[19]

The flexibility of these settlements manifests itself as highly adaptive business connections, social organizations, forms of life, and building practices. This chapter focuses on the economic activities that developed in the low-income quarters of Nairobi—areas where at least one-third of the total urban population was living in the 1970s.[20] The cases that follow show how inhabitants developed innovative solutions to the challenges they faced. Without romanticizing the situation in the flexible settlements, I bring out the ingenuity of the people living there. Rather than passive

[19] Di Nunzio, Marco, "'We are Good at Surviving': Street Hustling in Addis Ababa's Inner City," *Urban Forum* 23 (4), 2012: 433–447, here: 441.

[20] Herrle, Peter, Henning Lübbe, and Jakob Rösel, *Slums und Squatter-Siedlungen: Thesen zur Stadtentwicklung und Stadtplanung in der dritten Welt*. Stuttgart: University of Stuttgart, 1981, p. 52.

victims of unfortunate circumstances, the people of Ndururu and Mathare Valley creatively and ingeniously made the world in which they lived. Rather than being isolated from the rest of the city, the so-called slum dwellers took active part in a partly global consumer culture and maintained ties far beyond the borders of the settlement itself. Inhabitants set up businesses, built their own homes, and organized the provision of basic amenities, including water.[21] In Mathare Valley, many projects were organized on a cooperative basis. We also find a couple of registered companies, including the Mathare United Traders and Farmers Company Limited, which tried to expand its operations beyond the borders of the Valley.[22]

To some extent, the chapter is my attempt to further illustrate the point that prominent economists Abhijit Banerjee and Esther Duflo made in their book *Poor Economics*.[23] Being "poor" does not mean being resigned and passive. On the contrary, Banerjee and Duflo argue that, when given access to, for example, microcredits and affordable insurance, many low-income individuals and collectives can better exercise their creativity and stabilize their businesses. By highlighting the inhabitants' flexibility and creativity, I align Ndururu with AbdouMaliq Simone's description of life in African cities: "African cities are characterized by incessantly flexible, mobile, and provisional intersections of residents that operate without clearly delineated notions of how the city is to be inhabited and used."[24] The dwellers we meet in this chapter have much in common with the "seemingly marginalized" Johannesburg inhabitants, whom Simone refers to as incessantly innovative—to a large extent because their daily lives have a temporary and constantly changing character.[25]

Like Banerjee and Duflo, I recognize that many low-income people "are energetic and resourceful and manage to make a lot out of very little."[26] As Banerjee and Duflo also point out, it is possible to improve the situation of the "poor" by concerted efforts such as improved education

[21] Akallah, Jethron Ayumbah, and Mikael Hård, "Under the Historian's Radar: Local Water Supply Practices in Nairobi, 1940–1980," *Water Alternatives* 13 (3), 2020: 886–901.

[22] *The Sunday Post* (Nairobi), March 9, 1969.

[23] Banerjee, Abhijit, and Esther Duflo, *Poor Economics: A Radical Rethinking of the Way to Fight Global Poverty*. New York: Public Affairs, 2011.

[24] Simone, AbdouMaliq, "People as Infrastructure: Intersecting Fragments in Johannesburg," *Public Culture* 16 (3), 2004: 407–429, here: 407.

[25] Ibid.

[26] Banerjee and Duflo, *Economics*, p. 218.

and medical support. My emphasis on creativity and flexibility should, thus, not be interpreted as an argument in favor of a laissez-faire society. The fact that I acknowledge settlement inhabitants' high degree of agency does not imply that I condone the State's abdication of its responsibility toward the underemployed and disenfranchised.[27] Rather, my argument is that we must acknowledge the creativity and innovativeness embodied by those "self-employed artisans, shoemakers, tailors, manufacturers of beers and spirits … musicians, launderers, shoeshiners [*sic*], barbers, night-soil removers, photographers, vehicle repair and other maintenance workers," whom Keith Hart encountered in the "informal sector" he investigated.[28] So-called slums cannot be reduced to decay and stench, crime, and illegality.

Making and Selling *Pombe* and *Chang'aa*

Ndururu—and the entire Mathare Valley—has a politically charged history. The British colonialists regarded both the Valley and the Pumwani settlement as hotspots for "nationalist political activity." During the so-called Emergency Period from 1952 to 1960, the colonial government became more aggressive in its surveillance of the areas and systematically carried out what it euphemistically labeled "slum clearance" policies.[29] Only in the early 1960s, when the state of emergency was lifted, could people return to Mathare Valley.

Less than a decade later, Ndururu was brimming with life. An estimated twenty-thousand people—roughly half of them children and adolescents—lived in the settlement. Among adult men, only five percent described themselves as unemployed.[30] According to a survey carried out in 1969 and 1970, the situation was different for women: "Two-thirds of the women called themselves housewives, thereby implying unemployment, but a large number of them probably brew beer, otherwise they would be destitute, and many of those on their own are prostitutes."[31]

[27] Cf. the famous analysis of the need to legalize the so-called informal economy in Peru by Hernando de Soto, *The Other Path: The Economic Answer to Terrorism*. New York: Basic Books, 1989.

[28] Hart, "Income," p. 69.

[29] Hake, *Metropolis*, p. 133.

[30] Etherton, *Mathare*, p. 39.

[31] Ibid.

The city authorities regarded beer-brewing and gin production as the main causes of social problems in Nairobi's squatter areas. These activities were both an opportunity and a threat. The brewing of millet beer—*pombe*—and the making of other, partly illicit, alcoholic beverages provided many dwellers with much-needed earnings, but it also gave the authorities good reasons to police the areas. Somewhat exaggeratedly, Hake suggests that the "economy of Mathare Valley was essentially a *pombe* economy, as more than half the people were dependent upon the brewing of traditional African beer."[32]

"Native beers" differed from European-style ale or lager beer, as they were not made of rye or wheat. As mentioned, *pombe* was usually made of millet. Another variety of native beer, *busaa*, was made from corn, sorghum, or millet. Often, beers contained some kind of sweetener. In the case of another variety, *muratina*, honey was added to support the fermentation process, as with the manufacturing of mead in Europe.

In addition to these beers, establishments in Mathare Valley offered several varieties of so-called Nubian gin. The generic name for the distilled products was (and still is) *chang'aa*, but they were also marketed with more evocative names, like *gatoogo* (Smoke) or *kang'arikia* (Kill Me Quickly).[33] Some people specialized in producing *munyeki*, made with water, honey, sugar, and yeast.[34]

Despite regular police raids, manufacturers of these illicit beverages apparently made decent profits, especially from sales of their products to customers beyond Mathare Valley. Soldiers from the nearby Air Force base were among the most frequent customers.

Beer and gin production may have been small scale, but it was an integral part of an enterprise that extended far beyond the settlement. Hake describes how, in the neighboring village of Gichagi, an economic system had evolved around *pombe* and *chang'aa*. Although brewing and distilling took place locally, the enterprise relied on supplies from the outside world. Settlement shop owners provided the manufacturers with brown sugar: "The demand for brown sugar is so great that shopkeepers order several bags at once. A whole bag of brown sugar can be cleared within a few hours." In addition, shopkeepers also kept necessary stocks of baking powder. Alternatively, brewers could also use molasses, a highly valued

[32] Hake, *Metropolis*, p. 157.
[33] Ibid., p. 158.
[34] Ibid., p. 115.

product which was "delivered by Land-Rover, with a guard mounted over the load."[35] To satisfy the tastes of *chang'aa* customers, distillers also required "Magadi Soda [sodium carbonate] in [the] form of white bars."[36] When brown sugar or molasses were not readily available, brewers could also use *munyeki* as a starter.

To make gin, the "ingredients are mixed to gether [*sic*] to form a solution which is emptied into tanks or drums buried along the banks of Nairobi River." After fermentation took place, the containers were unburied and heated, to start the distillation process. The people of Mathare Valley opted not to produce liquor in the village proper, because it was dangerous:

> Distillation of Chang'aa can be a disastrous activity. Many precautions have to be taken to ensure that the container's [*sic*] can withstand the great pressure. If the containers burst they set free the infuriated contents which can easily claim a man's life or set a whole village on fire.

Another reason for working with the drums on the riverbank was the easy access to water: river water was used to cool the containers after distillation.[37]

Brewers and distillers ran further risks. Since very few of them had production licenses, they were being regularly "haunted [*sic*] by the Police." Perhaps producers felt relieved each time their ready-made product passed from their possession to local wholesalers. In turn, wholesalers sold the beer or spirits to barkeepers and retailers. Then it was the resellers who clashed with the law—and were forced to spend considerable sums to "quieten the roving Policeman."[38] Witnesses from the Pamwani area indicate that paying off police was a successful strategy most of the time—the police left customers alone:

> All during the week, workers filter in from different parts of Nairobi to sit on long benches next to the mud houses and drink the finished product out of Kimbo lard cans that have handles neatly soldered on them. However, Saturday and Sunday are the big days, and the village is crowded with customers of many tribes, attempting to dance to traditional music played by

[35] Ibid., p. 108.
[36] SOAS, Hake, PP MS 46: File No. 4:31, "Gichagi Housing Development."
[37] The descriptions in the last two paragraphs are based on ibid.
[38] The two quotes in this paragraph are taken from ibid.

old men, eating meat in the makeshift restaurants, or just staggering around after "dropping too many Kimbos."[39]

These customers and patrons were end-users in the economic chain. Another end-user group was farmers, who bought the brewers' and distillers' waste grains to use as fertilizer for their fields. This is further evidence that flexible settlements were linked to the wider world on both ends of the beer- and gin-production cycle.

The *pombe* and *chang'aa* case illustrates the ways in which settlement dwellers adapted to their circumstances. The case also confirms the existence of economic ties between the low-income flexible settlements and the more established parts of the city. The beer-and-spirits example goes a long way toward challenging the conflation of "informal" commercial activities with "illegal" ones. It was the municipal authorities' prohibition against producing and selling "native beer" and Nubian gin in Mathare Valley—as well as in other "informal" settlements—that perpetuated the idea of "slums" as places where crime was rampant. A more convincing counter-narrative is that it was the authorities themselves who defined the "shantytowns" in terms of informality—by declaring some of the inhabitants' commercial behavior to be illegal.

Andrew Hake's observations support the latter conclusion. In a sociological survey of Mathare Valley conducted in 1967, Hake and his co-author Marc Howard Ross specify that breweries were not illegal in all parts of Nairobi:

> If it were, the neighbouring Ruiruaka Valley, which is the centre of Kenya's modern brewing industry, would be a target for closure. In any case, "native beer" is already brewed under licence in Pumwani and elsewhere, and is available in bars throughout the Eastlands area.[40]

Initiating the "transitional economy" concept, Hake and Ross argued that the City Council ought to support rather than criminalize economic activities in the squatter areas. Employment in "modern" industries was not a viable alternative for the majority of uneducated and largely illiterate dwellers. Instead of focusing solely on large-scale production plants, the government needed to develop an economic policy that recognized the

[39] McVicar, "Twilight," p. 82.
[40] SOAS, Hake, PP MS 46: File No. 4:17, "Mathare Valley Housing Project."

intrinsic potential of the unplanned settlements. To this end, the authorities would have to include what Hake and Ross called "home industries" in their political programs.[41]

Another Contested Commodity: Meat

In a 1969 survey, Joan Richard, another of Hake's colleagues at the National Christian Council of Kenya, makes a similar observation about the meat trade. A close look at this business in Nairobi supports the argument that the City Council imposed a double standard on commercial ventures, with draconian rules for the flexible settlements, and more lenient measures for the city's wealthier areas. Richard claims that the authorities, by prohibiting small food retailers in Mathare to sell meat, forced many shop owners to buy wholesale meat illegally at high prices from "big business men." Voicing the concerns of retailers and inhabitants, Richard records that "the people at the Mathare Valley see these [strict license policies] as nothing but sheer exploitation."[42]

In Mathare Valley, some fifty shops provided the community with soft drinks, vegetables, and other products: "In the shops you find all the commodities [just as] in any other big shop everywhere." As with brown sugar, shopkeepers received provisions from Greater Nairobi and the surrounding countryside. Despite the settlement's purported informality, two shops carried official licenses, allowing them "to get their meat from the Kenya Meat Commission." This meant that the meat was monitored and regulated by veterinarians and other medical experts, according to a legal framework. The fact of settlement shops being licensed to buy meat also shows that city authorities did not deem all activities in "slums" to be illegal.

Documents from the Nairobi Social Services and Housing Committee indicate that the city indeed issued official licenses to small businesses in "informal" areas. In one case, when a Pumwani bar owner shuttered his business, the city offered his license to other interested entrepreneurs.[43] Another survey conducted by researchers at Nairobi University College in

[41] The quote and the material in this paragraph are based on ibid.

[42] Ibid.

[43] The material in this paragraph is based on ibid. and SOAS, Hake, PP MS 46: File No. 4:14, "Kariobangi Estate."

1969 reveals that as many as one-third of the shopkeepers in the notorious Kariobangi area held official licenses.

According to Richard, the official explanation for the harsh licensing policy in flexible settlements was not necessarily based on the shops' location in "informal" parts of town. Richard reports the alleged reason, as provided by an employee of the Nairobi City Town Planning Office: the majority of shops in Mathare Valley were "too small and situated in unhealthy conditions." Given the restrictions, censured shop owners had to choose between procuring meat "at a price" from "big dealers" or resorting to "the black market." This strategy illustrates the flexibility of their business networks—and flexibility emerges as a key feature of underground markets.[44]

The Shift from Cooperative to Company Status

The beer, gin, and meat examples show that the authorities treated low-income neighborhoods in a biased manner. On a regular basis, the Nairobi City Council authorized Nairobi's police force to round up beer and liquor manufacturers as well as tax evaders and other alleged lawbreakers.[45] Apparently, some among the general public shared this negative view of "slums." John Abuoga, a columnist for *The Sunday Post*, observed in 1969 that vocal critics "are quick to compare it [Mathare Valley] with Chicago for its association with the illcit [*sic*] liquor industry and for the hospitality it extends to some of the undesirable elements of Nairobi's society." Critics also accused the Valley of being "an eyesore and a serious hazard to the health of the capital." Some even urged the authorities to demolish the entire settlement.[46]

Abuoga took a more conciliatory stance. Impressed by the flourishing businesses and the responsible village committees, he argued that Mathare Valley had to be "recognised as an established settlement." Abuoga called attention to the many "shops, stores, laundries, butcheries, bars and restaurants" in the area, and he praised the organization of the five main villages. Headed by local chairmen—including Ndururu Kiboro in Village

[44] The material in this paragraph is based on SOAS, Hake, PP MS 46: File No. 4:17, "Mathare Valley Housing Project" and SOAS, Hake, PP MS 46: File No. 4:48, "Nairobi Industrial Area."

[45] Hake, *Metropolis*, p. 160.

[46] The quotes in this paragraph are taken from *The Sunday Post* (Nairobi), March 9, 1969.

II—the village committees mediated in cases of social conflict. These committees also championed new, collaborative, technical- and social-infrastructure projects: water pipes and centrally placed latrines; nursery schools and kindergartens; communal and cooperative buildings. With great admiration, Abuoga commended "the campaign to raise funds for a Harambee secondary school, a project initiated by the Member of Parliament for the area, Dr. Munyua Waiyaki."[47]

Usually translated as "let's pull together," the concept of *harambee* had become an important policy feature of the ruling party—the Kenya African National Union (KANU). Shortly after Independence, Jomo Kenyatta, the young nation's first prime minister, launched the *harambee* campaign. It was an effort to mobilize the populace to contribute to achieving (allegedly) common goals. The government often used *harambee* in reference to educational institutions. *Harambee* schools were usually co-funded or co-constructed: if the local community reserved land for a school and contributed to its construction, the government would commit to buying educational materials and to pay teachers' salaries.[48] Settlement dwellers could choose to contribute a small sum of money, some building material, or their own labor. Local authorities in Mathare Valley were satisfied if inhabitants were "willing to contribute one nine inch stone or Sh.2 [2 Kenyan shillings] each."[49]

Abuoga used the word *harambee* only in relation to the officially promoted secondary-school project. In other cases, he preferred to talk about "self-help" initiatives, "joint efforts," "self-built" buildings, and "co-operative" arrangements. "Self-help" always carried a collaborative connotation, whereas "home-made" referred, for example, to the construction of "flimsy buildings with tin roofs and roofs of cardboard cartons or crate timber," which individuals put together to house their families or their businesses. The efficacy of cooperative effort came out strongly in the case of technical infrastructure. Cooperatives made sure that parts of Mathare Valley were connected to the centralized, municipal water supply system. Cooperatives also ensured that electricity was installed in the main social halls. And Abuoga noted with astonishment that one could find television sets in some assembly halls.

[47] The quotes in this paragraph are taken from ibid.

[48] Hill, Martin J. D., *The Harambee Movement in Kenya: Self-help, Development and Education among the Kamba of Kitui District*. London: The Athlone Press, 1991, p. 48.

[49] Hake, *Metropolis*, p. 157.

Concepts like "self-help" and *harambee* were not meant to imply total autonomy and independence on the part of the villages. On the contrary, the intention was for State bodies or other organizations to assist local actors. Employing the concept of "aided self-help," government reports celebrated the administrative, financial, and technical support that various communities received from the National Housing Corporation (NHC). The National Christian Council of Kenya (NCCK) was a nongovernmental organization (NGO) particularly active in Nairobi's low-income settlements. Financially supported by North American and European churches, the NCCK in 1969 created the Mathare Valley Project. This initiative, headed by an advisory committee, coordinated activities to improve inhabitants' conditions. Surpassing traditional charity initiatives, the efforts included establishing schools and "home industries," as well as organizing teaching and garbage collection. "Visitors and residents of the villages joined in the construction of schools, offices and small-industry workshops."[50]

In Mathare Valley Village I, the NCCK's home industry program helped to found a needlework cooperative, which soon turned a profit by selling puppets and dolls. The cooperative even made it onto national television, appearing on "Art for the Young," a program supported by the Ministry of Education. Perhaps surprisingly, the "puppetry coop" included women as well as men, and co-op members used not only traditional methods, but experimented with "new production techniques such as molds." Their product line included zebras, giraffes, and crocodiles. Yet another group, partly supported by the German Volunteer Service, dedicated itself to making wooden trucks, cars, and other toys.[51]

The NCCK home industry program reflected the cultural values of European and North American Protestant churches. Nevertheless, the existence of this and other cooperative initiatives support the observation that the Mathare villages comprised an active, versatile, and flexible settlement. The Nairobi City Council officially acknowledged the home industries cultivated by the Mathare Valley Project—evidence that the government's approach to "shanty towns" was somewhat ambivalent.

[50] The information and quotes in this paragraph are taken from SOAS, Hake, PP MS File No. 4:8, "Kenyan Ministry of Housing Reports," as well as PP MS 46: Files No. 4:17 and 4:18, both entitled "Mathare Valley Housing Project."

[51] The information and quotes in this paragraph are taken from SOAS, Hake, PP MS 46: Files No. 4:17 and 4:18, both entitled "Mathare Valley Housing Project," in addition to SOAS, Hake, PP MS 46: Files No. 4:26, "Slums/Squatters/Shanty Towns."

Importantly, many project initiatives originated with the people of Mathare Valley themselves, rather than from outside the settlement. In 1965, village women founded the "Women's Progressive" (*Mandeleo y Wanawake*) as a production cooperative. They began by manufacturing "*kitenge* shirts in a small room in Village II."[52] Production soon diversified to include crocheting, knitting, weaving, basketry and mat-making. Hake noted that the cooperative also acquired "three sewing machines, but doesn't know how to use them!" Despite working independently most of the time, all members profited from the security offered by the collective. For example, lower-income members of the cooperative were eligible for economic support to send their children to school. The cooperative structure also enabled the members to accept larger orders, for example to make school uniforms.[53]

Cooperative action took on even greater importance when it came to land use and building construction. In the late 1960s, a policy change quickly gave rise to a large number of local initiatives; Mathare Valley was a hotspot for collective action and self-help housing—if only briefly. The largest cooperative society counted nearly five hundred members. The Ministry of Housing, the Nairobi City Council, and the NCCK joined forces to support the cooperatives and companies—both financially and technologically.[54] Although the municipality was obliged to supply villages with water, to lay sewerage pipes, and to construct roads, the City Council seldom delivered the necessary services. The inhabitants were compelled to provide for themselves, which meant that supplies were distributed unevenly. The majority of cooperatives and companies had access to running water and a pit latrine somewhere on the premises. Elsewhere, however, access to water ranged from one tap per hundred residents to a single tap per six thousand people. The city collected garbage on a regular basis for only one in five households. Electricity was absent in the residential areas. Clearly, the KANU State did not fulfil its obligations.[55]

Despite the City Council's only half-hearted assistance, Mathare Valley continued to attract people from upcountry areas of Kenya. Because of skyrocketing costs and increasing market orientation, the character of the

[52] Hake, *Metropolis*, p. 157.

[53] The quotes in this paragraph are taken from SOAS, Hake, PP MS 46: Files No. 4:17 and 4:18, both entitled "Mathare Valley Housing Project."

[54] Hake, *Metropolis*, p. 168.

[55] Etherton, *Mathare*, p. 49.

cooperative movement quickly changed. When cooperative members realized that declaring company status would enable them to sell their individual plots or shares to outsiders at a profit, many cooperatives were transformed into companies. This move helped to simplify access to capital and to turn collectives into shareholder groups. Speculation replaced self-help. The outcome was feverish construction activities and a rapid population increase. Within less than two years, the valley's population doubled. Owners began to sublet their apartments—and earn staggering profits.

In 1970, fourteen housing cooperatives and companies and were officially registered in Mathare Valley. One of the official co-ops was the aforementioned United Traders and Farmers Company. This cooperative bought parts of Village I, in addition to some uninhabited land where it built "timber houses with corrugated iron roofs" for some of its two hundred shareholders.[56] One of the companies was the Nyandarua Youth Wing and Company, which provided housing and ran a small bar with the unlikely name "The Alaska Hotel."

The shift from cooperative to company carried a certain degree of professionalization. Boards of directors and various committees made strategic decisions and employed managers to implement those decisions. Increasingly in need of expert knowledge and service personnel, companies hired "clerks, advocates, surveyors, draughtsmen and contractors and other technical aid on a permanent or temporary basis." One such company employed "three cleaners for the latrines and to collect refuse; and six nightwatchmen whose duties are to guard the company offices and the housing."[57]

Absentee ownership was commonplace. A structure called Company No. 6 exemplified this, as only three of the firm's 160 members actually lived in Mathare Valley. Tenants were willing to pay for a comparatively high standard of living. The density of latrines was high, and some dwellers enjoyed the luxury of access to individual kitchens, simple bath cabins and collective washrooms. The company had designed concrete channels to guarantee efficient stormwater drainage. The Nairobi City Council made sure that garbage was collected regularly on the site.[58]

In June of 1970, Kenya's Minister of Housing, Mr. Paul Ngei, visited Mathare Valley. During the trip, Ngei promised to commit considerable

[56] The information and quotes in this paragraph derive from Hake, *Metropolis*, pp. 165–166.

[57] Etherton, *Mathare*, pp. 46, 51.

[58] Ibid., p. 51.

funds to support building activities in the settlement. This alone testifies to the sharp changes in official policy. A couple of months later, the local authorities declared their willingness to guarantee cooperatives and companies official titles to their lots.

I contend that, by the early 1970s, Mathare Valley had ceased to be an "informal" settlement—if it had fit that description in the first place. By registering cooperatives and companies, the authorities officially recognized their existence and rights; by investing large sums in infrastructure, authorities acknowledged that the "shanty towns" had come to stay.

The Paradox of the "Uncontrolled Settlement"

Although it is a misnomer, the term informal settlement is still frequently used to refer to Mathare Valley. Another misapplied concept—this one historic—was that of the "uncontrolled" settlement. This reference appeared in the subtitle of a detailed study conducted in 1969–1970 by architects and sociologists at the University of Nairobi.[59] The full title of the study was *Mathare Valley: A Case Study of Uncontrolled Settlement in Nairobi*, and it was published in 1971 by the university's Housing Research and Development Unit. David Etherton was the main author and editor. The study is invaluable to scholars: it documents living conditions in the valley as official archives seldom do. Based on participatory observation and many interviews, Etherton's investigation offers insight into the Valley's material structures, both residences and infrastructure.

Consider Etherton's description of Village I. The great majority of buildings in this settlement were made of mud and wattle. Construction was swift and fairly simple; the outcome was buildings defined by Etherton as "semi-permanent." With "enough friends" and enough construction material, the builder could complete a four-room house within one week. Most builders used eucalyptus trunks and branches to form a sturdy frame. First, poles were sunk vertically into the ground; then, strips were tied or nailed horizontally to the poles. Rubble from the nearby quarry may have been used as bulk filler material. To stabilize the structure, interior and exterior walls were plastered with mud or cow dung. Many builders fitted cardboard to the outer walls. To protect their buildings from heavy rains, people covered their roofs with "second-hand corrugated iron, flattened tins, and flat sheets." The lowest-income dwellers resorted to using scrap

[59] Etherton, *Mathare*.

material. Floors of tamped-down soil were made easily, while the construction of windows and doors required special skills. To guarantee a certain degree of security, carpenters or *mafundi* (technicians) designed wooden doors and door frames as well as closable window shutters. Clearly, despite the collaborative nature of the construction process, individuals, cooperatives, and companies relied on expert knowledge:

> It is difficult to make a monetary estimate of labour involved in the construction of squatter housing because so much of it is built on the basis of mutual self-help. Clearing the site, digging holes for the posts, and building the walls is often done by the women but if a fairly substantial roof is built with, say, corrugated iron, somebody with the necessary skill (and tools) might be called in and paid for his help.[60]

As with other economic activities, construction required supplies from beyond the Valley:

> Gum-poles [poles made from the gum tree] are bought in large quantities from the Forestry Department in Karura or from small firms in Central Province. Poles are also supplied by agents in Ngara and other parts of Nairobi who sell a wide variety of building material including second-hand timber, flattened metal, drums etc.
>
> More portable materials such as cardboard, paper and polythene are often collected from the industrial area by individuals and then sold in the valley at a good profit.[61]

These contemporary quotes reveal the flexible settlement's deep ties to the formal—rather than the informal—economy, including public institutions. The diversity of building materials also indicates the adaptability of builders in these low-income settlements. The design of houses suggests that inhabitants valued flexible solutions: when faced with eviction by the authorities, dwellers could dismantle their eucalyptus skeleton-structure with relative ease.

Beds, too, were constructed with an eye toward being dismantled: carpenters used pegs to hold together the wooden bedframe. This allowed the bed to be easily taken apart, transported, and reassembled at a new

[60] All quotes in this paragraph are taken from ibid., p. 24.

[61] Ibid. Detailed cost estimates may be found in SOAS, Hake, PP MS 46: Files No. 4:33, "Kitui Village, Pumwani."

site, if necessary. Most mattresses were filled with sisal, a plant extensively cultivated throughout Kenya. The bed base was usually made of "interlaced rubber strips cut from discarded car tyres," which were "extensively used in Nairobi for the purpose."[62] While only very few people in Mathare Valley owned an automobile, the use of car tires shows involvement nonetheless in the burgeoning, international car economy.

Etherton also documents how typical families furnished their homes: a table, a sofa, several chairs, a charcoal cooking stove (*jiko)*, and one or two beds (an average of 2.5 people per bed). In some villages, the research team even found houses with radios. To store personal belongings as well as food, residents used suitcases, boxes, baskets, and jars. Given that houses did not have running water, residents used four-gallon cans (18-liter *debes*) to retrieve water and 44-gallon (200-liter) oil drums to store it. About half of the dwellings had shelves or cupboards. Dwellers employed aluminum pots (*sufurias*) for cooking; for eating, they used mugs and plates made of enameled tin, "sold in the market and by hawkers." In most homes, people also used iron bowls (*karais*) for personal hygiene and to wash clothes. Only a minority owned either kettles or frying pans. Conforming with tradition, people ate with their hands rather than with forks.[63]

Aluminum pots, iron bowls, oil drums, tire rubber: the use of all these implements and materials proves that people in the flexible settlement took an active part in the global economy.[64] Mathare Valley dwellers also participated in the global economy when it came to lighting their homes. Etherton found that each room was equipped with a "wick lamp using kerosene as fuel"; kerosene is a product of the petroleum industry. People in the settlement devised a neologism for the oil lamp, nicknaming it *nyitira njare*, a Kikuyu expression meaning "hold it for me while I make the bed."[65] The oil lamps were also "one the many products of the local 're-use' industry." The people of Mathare Valley and the immediate surroundings practiced "recycling" decades before the concept debuted in political debate. The same holds for the manufacturing of *jikos* [charcoal

[62] Etherton, *Mathare*, p. 37.

[63] The information and quotes in this paragraph are taken from Etherton, *Mathare*, pp. 37–38.

[64] For a more extensive chronology of the role played by African producers and consumers in the global economy, cf. Prestholdt, Jeremy, *Domesticating the World: African Consumerism and the Genealogies of Globalization*. Berkeley, CA: University of California Press, 2008.

[65] Kikuyu was the largest of the ethnic groups living in Mathare Valley.

cooking stoves], which were "produced exclusively by African artisans from scrap metal."[66]

Charcoal and piped-in water were for sale at designated locations in the villages. Inhabitants purchased kerosene and other daily necessities in local shops. These small stores offered many commonly used food items: potatoes, beans, cabbage, fruits, onions, yams, and other produce. Stores also carried cornmeal, meat, eggs, milk, bread, as well as soft drinks and beer.[67] Perhaps unexpectedly, in more than half of the homes surveyed, the research team found that meat was prepared: mutton and chevon, beef and pork. To offset the relatively high prices of such products, dwellers often opted to fill their stomachs with *ugali* (boiled cornmeal) and *sukuma wiki* (literally, "pushing a week"), a variety of cabbage so inexpensive that even the poor could afford to eat it every day.

The designations "uncontrolled" and "informal" represent a bird's-eye view of Mathare Valley, a view employed by the paternalist social scientist or the uneasy urban planner. If we look at so-called slums from a micro perspective, rather than from above, we see another world. This world is inhabited by people who carry out ordinary, daily routines: working and studying, cooking and eating, sleeping and enjoying themselves. As the Nairobi microhistory shows, these ordinary people invested—albeit modestly—in housing and furniture, and they were well integrated in the economy beyond the settlement.

"Self-help" and "Site-and-Services"

In early 1970, *askari* (military forces) set fire to the houses in the Nairobi "shanty town" of Kaburini, leaving roughly 1,700 people homeless and bereft of their belongings. The Minister for Local Government denounced the act of arson, and the press followed suit. University students responded

[66] The quotes and information in this paragraph are taken from Etherton, *Mathare*, pp. 37–38.

[67] Visitors to the Pumwani low-income settlement observed shops carrying the following items: "loaves of bread, cakes, barns sweets of different kind in glass jars. Soap, Matches, Cigarates, Combs, Carry powder, fanta, handkerchiefs, safety pins, looking glasses, small brushes, shoe-laces, plates, cups, cooking oil, lanterns, small basins, sugar, salt, maize flóur, parrafin, razor blades, pencils, small exercise books, ink, onions, tea-leaves, needles, and threads, small source pans, table spoons, and many other articles" [original spellings]; SOAS, Hake, PP MS 46: File No. 4:23, "Pumwani Housing Estate."

to the local minister's denunciation by demanding, "Give us your Mercedes to live in!"[68]

Minister of Housing Ngei's aforementioned visit to Mathare Valley was a direct reaction to this disastrous event. It signaled the KANU government's growing concern about problems in Nairobi's "squatter" areas. Exposing tensions between the national and municipal administrations, the minister addressed the harsh policies imposed the Nairobi City Council on allegedly illegal settlements. The minister also addressed health hazards and urban eyesores. Announcing new housing programs on the spot, Ngei criticized indirectly the city's biased stance on low-income neighborhoods. Ngei underscored the State government's concerns by offering a round of funding to build new houses and infrastructures in Mathare Valley and other communities in need.

The State government had previously pledged housing programs, but only now were more substantial funds promised. Under pressure from both the national government and the public, the City Council realized it had to comply. In September of 1970, the Director of Social Services and Housing announced that more than ten million Kenyan Shillings would be invested in the Valley alone.[69] That same year, the National Housing Corporation built as many as 1,300 houses in various parts of the city, though the rents were higher than the vast majority of people could afford.[70]

Aid to "squatters" and the homeless took various forms, including "site-and-services" schemes. According to this model, the authorities outfitted a designated area with basic amenities: a few main roads, one or two water taps, latrines, surface drains. In some cases, the government even provided street lighting and electricity for key buildings, as well as sewage pipes for the community. This was technical infrastructure only; inhabitants were largely left to their own devices when building houses, though sometimes the city supplied building materials.

Site-and-services schemes predated independence. In mid-1963, for example, the *East African Standard* heralded the beginnings of one such project in the Kariobangi area. One of the newspaper's articles, titled "Do-It-Yourself Solution for Squatters," reported that the Nairobi City

[68] The quotes and the information in this paragraph are taken from Hake, *Metropolis*, pp. 121–123.

[69] The information in this paragraph is taken from ibid., pp. 168–169.

[70] Ibid., p. 85.

Council had commissioned the site survey, "supplied water and roads," and provided people with building materials—though they were left to design and construct their own houses. Both the Minister of Local Government and the Minister of Labour and Social Services attended the public ceremony celebrating the official start of the project.[71]

Public-housing programs incurred various problems. For example, most of the people for whom the Kariobangi scheme was created never moved into the new houses. Instead, they decided to live elsewhere and to sublet rooms, or to sell their plots or houses to buyers of greater means.[72]

Excessively strict construction regulations presented another hurdle. Residents who were destitute could hardly be expected to build their homes out of brick or stone; nor could they afford corrugated-iron roofs. In 1950, Nairobi's previous mayor, Sir Ernest Vasey, had already recognized the issue—and called for greater "flexibility" of building regulations and standards.[73] In the 1960s, timber had proven to be a viable alternative to wattle-and-daub (interwoven sticks and twigs covered with mud or clay). Despite their objections that wood was neither fireproof nor very inexpensive, city authorities appointed a Timber Development Committee to explore the possibilities. As usual, the authorities' hypothetical inquiry into the potential of wood barely influenced building practices on the ground, and most dwellers continued to employ wattle-and-daub.

The National Housing Corporation's attempts at aid, including its various site-and-services schemes, failed to accommodate an urban population that was growing at an annual rate of six percent. Similarly, projects supported by the National Christian Council of Kenya and other NGOs fell short in scope. Even the Nairobi City Council's 1,200 plots in Kariobangi were a mere drop in the bucket. Once again, the people in settlements had no choice but to fend for themselves.

And they did fend for themselves. Less than one decade after independence, both Etherton and Hake emphasized the creativity of the Africans. Although the coming of formal independence did not bring total self-determination, the African population was about to learn how to best make use of its own resources. In the construction business, "self-help" was a necessary complement to capital-intensive, prefabricated mass housing. "The traditional building skills handed down by African families" still

[71] *East African Standard* (Nairobi), Aug. 10, 1963.
[72] Ibid., March 20, 1969.
[73] SOAS, Hake, PP MS 46: Files No. 4:2, "Vasey Report on Housing."

had a place, even in the modern era. According to Etherton, advanced, appropriate, intermediate, and traditional technologies are not "mutually exclusive."[74]

The Paradox of the "Transitional Urban Sector"

In his book *African Metropolis: Nairobi's Self-help City*, Hake expresses his belief that the people of Nairobi have the potential to help themselves. Despite Hake's vocation as a Protestant clergyman, his book was not written from an explicitly religious point of view: *African Metropolis* was meant to spotlight the economic and cultural power of people living in precarious, substandard conditions. In Hake's vision, it was the low-income dwellers themselves who comprised the self-help city: people from the settlements formed a "movement of urban development 'from below.'" This movement encompassed all aspects of life:

> The self-help city ... presents an alternative society to that of the modern city. It is an oral culture, relying less upon the written word than upon face-to-face communication. It is a culture in which women have a much greater role than in the male-dominated colonial pattern, and it provides, for the city's outsiders, an opportunity for integration.[75]

Hake had expressed these ideas in a talk given at the University of Sussex in 1970. He contrasted the modern city, with its office buildings, factories, and infrastructure "based on modern science and technology" with what he called the "*transitional urban sector.*" This swath of society is more than its "transitional economy"; the transitional urban sector is an entire "sub-culture" that "provides its own houses, jobs and culture." Hake describes this subculture as powerful enough to set off a "second urban revolution" that will transform the fabric of urban life as we know it. It would be a revolution from below.[76]

Hake addresses issues of trade and manufacturing at length. The economic activities of "shanty town" dwellers evince obvious parallels to the burgeoning trend toward "intermediate or appropriate technology," which I discussed in Chap. 4. Both approaches are feasible without large

[74] Etherton, *Mathare*, p. 78.

[75] This and the previous quote come from Hake, *Metropolis*, p. 172.

[76] SOAS, PP MS 46: File No. 4:45, "African Housing Policy." Large parts of Hake's 1970 talk later made its way into Ch. 11 of his book *African Metropolis*.

capital investments, and both include people without advanced vocational training. Three years before British economist E.F. Schumacher published his famed book *Small is Beautiful*, Hake made a case for intermediate and appropriate developments.[77] Archival material reveals that in the late 1960s, Hake was already corresponding with Schumacher's Intermediate Technology Development Group in London.[78] To paraphrase the subtitle of Schumacher's famous book *Small is Beautiful*, Hake envisioned an economy where "people mattered."

Hake's approach to what he called the second urban revolution was somewhat paradoxical: the majority of the population—low-income people—supposedly had potential to become an independent revolutionary force. Yet Hake and his church cohort actively sought to assist this ostensibly powerful majority. A project called the Small Industries Scheme is a case in point. In 1965, the Christian Industrial Training Centre, located in a Pumwani neighborhood, had created a committee to research ways of training and supporting would-be entrepreneurs. Under Hake's chairmanship, the committee discussed how to organize and finance a training center for this purpose. The committee also conducted several surveys, to help identify commercially promising products. In retrospect, the committee's proposed-product list looks anything but revolutionary; it included "Toys," "Christmas crackers," "Trays," "Burglar alarms," and "Book Binding."[79]

The small-industries committee used a report by the Reverend Charles R. Tett as its point of departure. Reflecting the immediate-post-independence need to "Africanize" the Kenyan economy, Tett suggested that "African-owned Small Industries ... should concentrate on using local skill & material to produce for the country many things now imported from overseas." According to Tett, the main problem was not necessarily a "lack of artisan skill"; it was that relatively few Africans had learned how to run a business—how to calculate costs, how to make a budget, how to secure raw materials, how to organize production, how to market and sell products. In a typically paternalistic manner, then, European and North American church representatives tried to impose capitalist ideas and methods onto the African craftsperson.[80]

[77] Schumacher, E.F., *Small is Beautiful: A Study of Economics as if People Mattered*. London: Abacus, 1973. The quote is taken from PP MS 46: File No. 4:45, "African Housing Policy."

[78] SOAS, PP MS 46: File No. 5:87, "Development of Small Industries."

[79] The information in this paragraph is based on ibid.

[80] The quotes and the information in this paragraph are taken from ibid.

From Informal to Flexible Settlements

The notion of the informal settlement has proven to be problematic. It relies on the assertion that many "slum" dwellers lacked formal titles to their land, and that their names were omitted from land registers. This "informality" did exist in Nairobi's low-income areas, but as we have seen, some dwellers in these districts did hold land titles. From the 1960s onward, the authorities registered several housing cooperatives and companies in Mathare Valley, and participants in the Langata site-and-services schemes also obtained formal titles.

By replacing "informal settlement" with "flexible settlement," we see the people of Mathare Valley from another angle: they were continually in the process of adapting, of inventing, of being *flexible* so as to survive.

Etherton viewed the people of Mathare Valley correspondingly. In his study of Mathare Valley, he frames the 1970s local economy as a "flexible system" in which craftspeople, shopkeepers, water vendors, and other providers of goods and services had learned to adapt to ever-emerging needs. Etherton writes, "The skills and techniques employed within the system are constantly changing to meet new demands."[81]

Geographer McVicar interpreted life in settlements in a similar way. During his fieldwork in Pumwani, McVicar encountered several people whose life histories illustrate this flexibility. One anonymous interviewee, "a big Kikuyu," told him:

> I'm only 45 but I've done a lot of things. When I first came to the city in 1934 I was a houseboy, then in 1939 I went over to Kariakor and became a mechanic. Finally I saved enough money to go into business, and in 1942 I got my own shop, selling general merchandise. [...]
>
> Finally I arrived here, and I still had half of the Shs. 600/, Shs. 300/ [600/300 Kenyan shillings]. I bought a laundry over behind the mosque—it was the kind of business that didn't take much capital. After one year I moved over here Now I have three people working for me part time and I generally make over Shs. 300/ a month myself. [...]
>
> I also have shares in City Breweries and in a hotel (café). [...]
>
> The hotel business is with my friends from Fort Hall, and I'm one of the directors. We meet once every month to discuss how the business is going.[82]

This personal narrative does not square with the "myths of informality" that Mike Davis references in *Planet of Slums*. The Kikuyu whom McVicar

[81] Etherton, *Mathare*, p. 78.
[82] MacVicar, "Twilight," pp. 203–204.

interviewed began his career in the city as a classical informal employee in the service sector, a member of what Davis calls the "informal proletariat."[83] But his story does not end there. Through savings and hard labor, the man was able to pursue some of his personal goals and start his own businesses. By adapting to the demands of the local economy, he exercised considerable flexibility.

Viewed from both local and global standpoints, we see that inhabitants' innovativeness and creativity allowed them to participate in a consumer culture that went far beyond the settlement's borders. People in Mathare Valley built houses with materials from the Nairobi area as well as other parts of Kenya. Aluminum, iron, and rubber came from even further away. In their daily lives, dwellers used *sufurias* [pots] made of aluminum, *karais* or *jikos* [stoves] made of iron, and *nyitira njare* [oil lamps] made of recycled materials. They slept in beds made by local carpenters who made use of rubber from car tires, and they stored their belongings in metal boxes and on wooden shelves. Etherton observed that "household utensils and furniture [were] all produced by African artisans using techniques which they… adapted to make use of second-hand materials."[84]

Like many of his contemporaries, Hake focused on the question of low-income housing. In doing so, Hake embraced many of the ideas in John F.C. Turner's *Housing by People*, a booklet published in 1976, a year before the publication of *African Metropolis*. References in *African Metropolis* show that Hake—like Etherton—was familiar with Turner's studies and that Turner inspired him to propose an economy that was open to all citizens. Turner called for the inclusion of the "poorest" communities in designing, constructing, and maintaining their own environment.[85] The more the authorities invest in large-scale housing programs, the worse the situation of the poor will be. Both Etherton and Hake subscribed to this conclusion. From the point of view of global history, it is important to note that Turner had lived in Peru for eight years, carrying out most of his research on housing policy in South America. Although inhabitants in Nairobi "slums" and Lima *barriadas* never met, their problems and possible solutions were discussed in a global network of critical architects and planners. But that is another story.

[83] Davis, *Planet*, p. 178.

[84] Etherton, *Mathare*, p. 78.

[85] Turner, John, F.C., *Housing by People: Toward Autonomy in Building Environments*. London: Marion Boyars, 1976.

CHAPTER 8

Confronting Menstruation in East Asia: Koreans Create Self-made Solutions

She experienced her first menstruation in 1960, and after she married, she used gauze as a sanitary pad. After ten years, she began to use several layers of gauze simultaneously. When the top piece got soiled, she removed it. Because the material was hard to obtain, she washed the pieces of gauze and reused them. She didn't have much material and used two or three pieces of the fabric per week.

The laundering method is worth mentioning, as well. She hid the used gauze in a dark corner, where no one would notice it, and she took it along with her dirty clothing when she did laundry. She rinsed and pre-washed the menstrual pads in running spring water, so the blood was washed away naturally by the current. Because she had been using the gauze for several days, it turned out to be impossible to remove all blood stains with this method,

M. Hård, *Microhistories of Technology*, Palgrave Studies in the History of Science and Technology,
https://doi.org/10.1007/978-3-031-22813-1_8

however. So, after rinsing the strips of gauze in running water, she soaked them in lye and pounded them clean.[1]

In this excerpt from an interview with a South Korean woman, the interviewer relates the recollections of sixty-seven-year-old Ohyeon-daek, as she thinks back to her early menstruation experiences. Ohyeon-daek was born in 1944 and had her first period at the age of seventeen. Given that most of her friends of the same age had already begun to menstruate, Ohyeon-daek was unsurprised when she saw blood in her underwear for the first time. At the night school she attended, Ohyeon-daek had already seen bloodstains on her classmates' clothing. In fact, for some time, she had worried about why her period had not started earlier. She knew that menstruation was a normal phenomenon, a sign of "sexual maturity"—and she was relieved when her first cycle began.

To prevent the menstrual blood from soiling her underwear and skirts, Ohyeon-daek initially received reusable menstrual pads, which her mother made from cotton fabric. Only ten years later did Ohyeon-daek adopt gauze pads, which she fastened to her underwear with a rubber band. While today, the use of recyclable pads may sound like a convenient, sustainable solution, Ohyeon-daek emphasizes the *in*convenience of this personal-hygiene technology. Given that she owned only a few of these pads, Ohyeon-daek was forced to use them for several days in a row—and sometimes even an entire week. This caused the patches of dried blood to scratch the sensitive areas around her vulva. In the summer heat especially, pads used for several days tended to produce an unpleasant odor. In general, Ohyeon-daek's memories of menstruation—and the relevant technologies—are rather unpleasant. She always lacked sufficient sanitary pads, and the pad-washing process was a nuisance. Further, given that menstruation was a taboo topic, Ohyeon-daek refrained from talking about

[1] This quotation is taken from Baek, Min Jeong, "Wŏlgyŏnggyŏnghŏm-ŭl t'onghae pon yŏsŏng chŏngch'essŏng-ŭi hyŏngsŏng-kwa pyŏnhwa-Andong-si Pungsan-eup Sosan maŭr-ŭl chungsim-ŭro" (월경경험을 통해 본 여성 정체성의 형성과 변화-안동시 풍산읍 소산마을을 중심으로) ["Creation and Changes of Women's Identity According to Menstruation Experiences—Focusing on Sosan Village in Pungsan-eup Andong-si"], *MA Thesis*, Andong: Andong National University, 2012, p. 58. Given that I have no proficiency in the Korean language, all translations from Korean to English in this chapter have been edited by Youngju Lee, MA, from Technical University of Darmstadt. Ms. Lee was kind enough to provide me with most of the literature and literary sources for this chapter; she also transcribed the interviews she carried out in 2018 with thirteen South Korean women.

her monthly problems—including her pain—with other women. At one point in the interview, she recalls the intense shame she felt when her mother-in-law caught her with blood on her clothing.[2]

Ohyeon-daek had grown up as the third of six siblings. The family lived in the small village of Bomun-myeon, in the central part of the country, which, in 1948, became South Korea. At the age of nineteen, upon marrying, she moved to Sosan-ri, a village approximately fifteen kilometers south of her home village, and roughly twenty kilometers west of the city of Andong. At age twenty-one, she bore her first child; three more would follow.[3]

This description of Ohyeon-daek's early menstruation experiences is taken from an interview conducted in 2011 by historical anthropologist Baek Min Jeong. Ohyeon-daek was the youngest of seventeen women (the oldest of whom was ninety) from Sosan-ri whom Baek interviewed for her master's thesis. The excerpt above represents Baek's account of Ohyeon-daek's experience.

Baek's interviews suggest that Ohyeon-daek's struggle to come to terms with her period was typical of Korean women of her generation. In fact, Ohyeon-daek's approach to her menstruation experience may well represent the experience of earlier—and later—generations, as well. According to a survey from 1970, about one-third of Korean girls in Seoul did not know anything about menstruation before their periods began.[4] Further, menstruation was a topic seldom mentioned, and menstrual blood was something to be hidden. In her role as a housewife, Ohyeon-daek was expected to carry out her daily chores without complaining, regardless of whether she had her period.[5] In addition, Ohyeon-daek was responsible for washing her sanitary pads discreetly if not secretly—without anyone else noticing.

In some ways, however, Ohyeon-daek differed from the majority of girls who came of age in mid-twentieth-century South Korea. She expected menarche (the first onset of menstruation), and she was prepared for it.

[2] The information in the last two paragraphs is taken from Baek, "Creation," pp. 36, 39, 42, 54, 58, 62, 73–74, 87, 95.

[3] Ibid., p. 29.

[4] Kim, Che-han, "Han'gukyŏsŏng-ŭi ch'ogyŏng-e kwanhan chosayŏn'gu" (韓國女性의 初經에 관한 調査硏究) [authorized title: "A Survey on Woman's Menarchal in Korea"], *Nonmunchip* (論文集) ["Collection of Treatises, Seoul National University of Education"] 3, 1970: 167–180.

[5] Baek, "Creation," p. 42.

Ohyeon-daek also understood the connection between menstruation and pregnancy. In contrast to Ohyeon-daek, Chilgok-daek, Baek's oldest interviewee, knew only that grown-up women experience bleeding—but she had no concept of the physiological processes behind this phenomenon; nor did Chilgok-daek comprehend that menstruation is a prerequisite to childbearing.[6] Although Chilgok-daek had heard about menstruation, the onset of her first menstrual cycle, at age seventeen, still shocked her: she "put her blood-stained underwear in the furnace and burned it."[7] Most of Baek's interviewees told similarly harrowing stories about their menstruation experiences.

Another historian, Youngju Lee, interviewed Korean women born in the 1950s and 1960s.[8] For these women, the occurrence of their first period was also a shocking, frightening event. In most cases, neither their mothers nor other female relatives had prepared them for this dramatic rite of passage into womanhood; they did not know how to cope with the first sight of blood—and the attendant problem of bloodstains on their clothing. Some girls tried to dry or congeal the blood in front of the fireplace; others drank alcohol in the hope that it would stop the bleeding. Some women thought they had been badly hurt and had sustained a serious internal injury. They understood that menstruation was not a topic to be discussed openly; this gave rise to feelings of shame. To avoid being scolded by relatives and teachers, early on girls learned to hide their menstruation pads and blood-stained strips of cloth; they washed their clothes in secret, at night—or during the day, in places where they would not be seen. Those who were most desperate even went so far as to bury their underwear in the ground.[9]

[6] Note that the suffix "-daek" indicates that a married woman's name refers to her home town or village. Thus, Ohyeon-daek and Chilgok-daek do not belong to the same family.

[7] Baek, "Creation," p. 31.

[8] Lee, Young Ju, Irhoeyong saengnidae-ŭi toip-kwa wŏlgyŏng-hanŭn mom-e taehan insik-ŭi pyŏnhwa – 1960-1980 nyŏndae han'guk-ŭi kyŏngu (일회용 생리대의 도입과 월경하는 몸에 대한 인식의 변화 - 1960-1980 년대 한국의 경우) [authorized translation: "The Introduction of Disposable Menstrual Products and the Change of Perception about the Menstrual Body, in 1960s–1980s Korea"], *M.Sc. Thesis*, Seoul: Seoul National University, 2018, p. 14: https://dcollection.snu.ac.kr/srch/srchDetail/000000152465 [accessed Aug. 22, 2022]. Substantial parts of this chapter are based on material from Lee's thesis.

[9] Baek, "Creation," pp. 31–37; cf. Lee, "Introduction," in particular interviewee No. 2 (anonymous, born in 1958).

To what extent do Ohyeon-daek and Chilgok-daek's narratives represent the feelings and beliefs of other women in South Korea? Were their feelings of shame—and the techniques they used to manage menstruation—typical?

Industrialization and the Taboo of Menstruation

Anthropologists and researchers in other disciplines have found taboos around menstruation throughout the world; historians throughout the centuries have documented negative associations with menstruating women. In a classic book titled *The Curse: A Cultural History of Menstruation*, scholars Janice Delaney, Mary Jane Lupton, and Emily Toth present a comprehensive discussion on the taboo of menstruation.[10] According to the authors, in most "native" cultures around the world, young girls are kept isolated for long stretches of time after their menarche, for example. The authors of *The Curse* talk about "Taboos of Exclusion."[11] In the Old Testament, Jehovah explains to Moses that a menstruating woman is unclean and must be separated from the rest of society for seven days; this notion of uncleanliness has persisted in many cultures. It is not merely the blood that must be hidden; the menstruating woman herself is also considered taboo.

Such taboos have been slow to fade. In the industrializing world of the nineteenth and early twentieth centuries, for example, the taboos around menstruation continued to present women with almost insurmountable problems. As more and more women entered the workforce and gained access to higher levels of education, they spent a larger portion of their time outside the home. The question of how to hide the evidence of their menstruation—let alone how to handle their menstrual flow during long hours at work or in school—became pressing problems for women in both North America and Europe.[12] Designed to alleviate the situation for menstruating women at work and in educational settings, legal measures and

[10] Delaney, Janice, Mary Jane Lupton, and Emily Toth, *The Curse: A Cultural History of Menstruation*. Urbana, IL: University of Illinois Press, 1988 (rev. ed.; orig. 1976).

[11] Ibid., Ch. 1. For a critical discussion about various taboos surrounding menstruation, see Buckley, Thomas, and Alma Gottlieb, "A Critical Appraisal of Theories of Menstrual Symbolism," in: idem., eds, *Blood Magic: An Anthropology of Menstruation*. Berkeley, CA: University of California Press, 1988: 3–50.

[12] Vostral, Sharra L., *Under Wraps: A History of Menstrual Hygiene Technology*. Lanham, MD: Rowman & Littlefield, 2008.

human-resources policies often proved counterproductive. Although meant to protect women, the era's "modern menstrual politics"—as the authors of *The Curse* call it—turned them into second-class citizens, and frequently relegated them to the category of temporarily disabled workers or students.[13] Clearly, legal and policy protections for women were ineffective.

Given the need to draw on women as part of the workforce, US and European manufacturers were spurred to devise products that would enable women to better manage their menstrual flow. A "technology" was urgently needed—a means by which menstruating women could spend hours away from home without having to worry about losing their jobs or finding themselves in awkward or embarrassing situations. The practical challenge was clear: outside the home, it was impossible to wash and dry any item used to absorb menstrual flow, whether that item was a cloth sanitary pad, an adult "diaper," or a sponge, for example. The optimal industrial solution was a disposable product. Fortunately for manufacturers, medical doctors and nurses recommended disposable products, citing hygienic concerns.[14]

The first commercial sanitary pads and belts were introduced on the US market in the late 1880s, and in 1896 the consumer-products giant Johnson & Johnson launched Lister's Towel. Made of gauze-covered cotton, this mass-produced product was designed to be thrown away after use. Twenty-five years later, Kimberly-Clark followed with "Kotex," a pad made of a cotton-like cellulose product called Cellucotton.[15]

In the United States, the disposable sanitary pad was by no means an immediate success. The price-point for disposable pads was one deterrent; taboos against frankly advertising the product in newspapers and magazines was another obstacle. Even if a woman had learned about Lister's Towels or Kotex from friends or relatives, it took some self-confidence to ask for these products at the local drugstore, or at a department store.[16]

[13] Delaney, Lupton, and Toth, *Curse*, Ch. 6.

[14] Farrell-Beck, Jane, and Laura Klosterman Kidd, "The Roles of Health Professionals in the Development and Dissemination of Women's Sanitary Products, 1880–1940," *The Journal of the History of Medicine and Allied Sciences* 51 (3), 1996: 325–352.

[15] Heinrich, Thomas, and Bob Batchelor, *Kotex, Kleenex, Huggies: Kimberly-Clark and the Consumer Revolution in American Business*. Columbus, OH: Ohio State University Press, 2004.

[16] Delaney, Lupton, and Toth, *Curse*, pp. 129–130.

The less-courageous could order both disposable and nondisposable menstruation products from mail-order companies.

Perhaps as an outcome of women's increased participation in political and economic life in the 1920s, some of these taboos were relaxed. For example, *Ladies' Home Journal*, *Good Housekeeping*, and other US print media began to publish advertisements for Kotex and other brands of disposable pads. Kotex's biggest competitor was Modess, a "sanitary napkin" manufactured by the Personal Products Corporation, a subsidiary of Johnson & Johnson. Both images and text reveal that these ads were geared toward White, middle-class and upper-middle-class women.[17]

In a 1929 ad, the manufacturer Kimberly-Clark extolls Kotex's "advantages": "it is disposable, just like tissue … easily, quickly; wrapped in soft, specially treated gauze." Rather than adopting potentially offensive words like "menstruation" or "blood," the ad uses a combination of euphemisms and quasi-scientific language: "science has found a solution to woman's oldest hygienic problem." In addition, the 1929 advertising copy implies that Kotex has solved the menstruation problem worldwide: "the hygienic habits of women have changed all over the world."[18] This statement was patently false: it would be forty years before Kotex entered the South Korean market.[19]

A US-Style Consumer Culture in South Korea?

In 1945, as the Second World War ended, the Korean peninsula was practically divided in half. The Soviet Union took control of the North, and the US Army installed a military administration in the South. In 1948, the Democratic People's Republic of Korea (North Korea) and the Republic of Korea (South Korea) were officially created. The Korean War (1950–1953) cemented this division between the communist North and the capitalist-leaning South.

The US military presence in South Korea remained strong until the early 1970s; this extended presence implied that the young republic would become heavily Americanized. Indeed, US administrators played a key

[17] Freidenfelds, Lara, *The Modern Period: Menstruation in Twentieth-century America*. Baltimore, MD: Johns Hopkins University Press, 2009, Ch. 4.

[18] The quotes in this and the previous paragraph appeared in *Ladies' Home Journal*, April 1929; quoted from Freidenfelds, *Period*, p. 122.

[19] Lee, "Introduction."

role in Americanizing the country's educational system, for example; rallying to the slogan "New Education" (*sae kyoyuk*), six years of elementary schooling became compulsory for both boys and girls. In addition, access to higher levels of education became less dependent on social status.[20]

In her highly regarded book *Irresistible Empire*, historian Victoria de Grazia documents the increasing US influence in Europe. Focusing on economic and cultural matters, de Grazia illustrates how, in the postwar period, Europeans—particularly in Western Europe—adopted lifestyles and consumer patterns from the United States. US manufacturers and retailers deliberately—and effectively—promoted the image of "America" as a cultural export. Advertisements and films contributed to the circulation of US ideals—and to the acceptance of US products. Consumer-products behemoth Procter & Gamble was just one of several US companies that used aggressive marketing methods to sell their products in Europe.[21]

De Grazia concludes that the image of the United States had such a strong impact on European citizens that they came to regard the United States as an "irresistible empire." US consumer culture was the vehicle for convincing Europeans to strive for "the American way of life." The question remains whether this process of Americanization was similar in the case of South Korea. To what extent were the people of South Korea tempted by US consumer products and cultural symbols? Focusing on technologies applied by Korean women to accommodate their monthly periods, I question whether consumers associated disposable sanitary pads and tampons with "America."

Historian Deok-ho Kim has analyzed the impact of the United States on Korean society and culture. In the early twentieth century, "America" was already successfully projecting a positive image in Korea. During the Japanese occupation (1910–1945) and thereafter, US influence increased dramatically. After the Second World War, many American products entered the country via so-called US Army and Air Force post exchange (PX) stores. Frequently, these products—from coffee and cigarettes, to cameras, cosmetics, and Coca-Cola—found their way onto the black market. Meanwhile, propaganda reels and Hollywood movies provided South

[20] Lee, Gilsang, et al., eds, *The History of Education in Korea: A Sourcebook*. Seongnam: The Academy of Korean Studies Press, 2015, pp. xii, 1.

[21] Grazia, Victoria de, *Irresistible Empire: America's Advance through Twentieth-century Europe*. Cambridge, MA: Belknap Press, 2005, pp. 416–425.

Koreans, including those in rural areas, with concrete images of what constituted the "American way of life." As income levels rose, people in South Korea also adopted more expensive symbols of modern consumer culture: television sets, refrigerators, and washing machines, for example.[22]

Economic statistics support the idea that South Korea was moving rapidly toward becoming a consumer society. In the 1960s and 1970s, as Gross Domestic Product increased by an average of nearly eight percent annually, private consumption tripled, in real terms.[23] In addition, consumers doubled the amount of money spent on entertainment, "beauty" products, and health products—including disposable sanitary pads. Indeed, imported products from the West, such as blue jeans, beer, and guitars, came to symbolize this purported consumer-oriented leisure culture.[24] The influence of American culture was also evident in the growing popularity of leisure sports, including bowling, golf, and tennis.[25]

The rise of South Korea's consumer culture was a thorn in the side of Park Chung-hee, the country's president from 1963 to 1979. From the start, the Park administration prioritized the manufacturing sector, particularly the export industry. To restrict consumption, the government levied high import tariffs and, in the 1970s, introduced value-added tax.[26] The government also increased the tax on luxury goods such as Western liquor, automobiles, television sets, and electric household appliances.[27] At every opportunity, Park criticized the emerging leisure culture, as well as the excessive consumption of goods deemed unnecessary. Preaching

[22] Kim, Deok-ho, "Han'guk-esŏŭi ilsangsaenghwal-kwa sobi-ŭi migukhwa munje" (한국에서의 일상 생활과 소비의 미국화 문제) ["Issues of Americanization in Everyday Life and Consumption in Korea"], in: Kim, Deok-ho, and Won Yong-jin, eds, *Amerik'anaijeisyŏn* (아메리카나이제이션) ["Americanization"], Seoul: P'urŭnyŏksa 2008: 121–158.

[23] Lee, Sang-rok, "1970 nyŏndae sobiŏkchejŏngch'aek-kwa sobimunhwa-ŭi ilsangjŏngch'ihak" (1970년대 소비억제정책과 소비문화의 일상정치학) [authorized translation: "The Consumption Control Policy and the Politics of Everyday Life of Consumer Culture in the 1970s"], *Yŏksamunjeyŏn'gu* (역사문제연구) ["Critical Studies on Modern Korean History"] 17 (1), 2013: 137–182, here: 149, 159; Yi, Jong-Hyun, *History of Korean Modern Retailing: Repressed Consumption and Retail Industry, Perceived Equality and Economic Growth*. Leiden: Brill, 2016, pp. 35–38.

[24] Lee, "Consumption," p. 170.

[25] Song, Eun Young, "1970 nyŏndae yŏgamunhwa-wa taejungsobi-ŭi chŏngch'i" (1970년대 여가문화와 대중소비의 정치) [authorized translation: "The Politics of Leisure Culture and Mass Consumption in 1970s"], *Hyŏndaemunhak-ŭi yŏn'gu* (현대문학의 연구) ["Journal of Korean Modern Literature"] 50, 2013: 39–72, here: 54.

[26] Yi, *History*, p. 108.

[27] Lee, "Consumption," p. 152.

frugality and the value of hard work, Park criticized young urbanites, for example. Park's logic: by spending their weekends "dancing go-go," young urbanites demoralized farmers and other diligent citizens.[28]

The Park government's economic policy was influenced by US economist Walt Whitman Rostow, who argued that "underdeveloped countries"—as South Korea and similar countries in the "Third World" in those days were called—must prioritize investment in heavy industry and the production of capital goods. Invoking Rostow's model of economic development, Park believed South Korea was not yet ready to enter the stage of mass consumption. In line with this thinking, the government continued to support traditional markets and sought to inhibit the expansion of department stores. Park's economic policy, then, suggests that South Korea was not as heavily Americanized as Western Europe. While many individuals in South Korea embraced US products and US cultural symbols, Park, apparently, did not find "America" to be "irresistible"—unlike, for example, the West German Chancellor Ludwig Erhard, the so-called father of the West German "economic wonder" of the 1950s and 1960s.[29] A liberal economist at heart, Erhard believed open markets would guarantee "welfare for everyone" and—as in the United States—subsequent access to automobiles and other consumer products.[30]

Techniques for Self-made Sanitary Pads

At least on the political level, then, South Korea was staving off the process of full-fledged Americanization in the 1960s and 1970s. However, to what extent were Korean women still tempted by American attitudes toward menstruation and the products marketed to menstruating women?

For this chapter, in addition to relying on interviews conducted by Korean scholars, I have sourced material from South Korean magazines and newspapers. The primary-source material includes advertisements, articles written by journalists and self-proclaimed experts, as well as letters to the editors written by women readers. At first glance, these sources

[28] This quotation is taken from a newspaper article by Park in 1972; quoted from ibid., p. 169.

[29] Nelson, Laura, "South Korean Consumer Nationalism: Women, Children, Credit, and Other Perils," in: Garon, Sheldon, and Patricia L. Maclachlan, eds, *The Ambivalent Consumer: Questioning Consumption in East Asia and the West*. Ithaca, NY: Cornell University Press, 2006: 188–207.

[30] Erhard, Ludwig, *Wohlstand für alle*. Düsseldorf: Econ-Verlag, 1957.

corroborate the hypothesis that women in South Korea tended to associate disposable menstruation products with "America." For example, in 1966, when the Mughunghwa Hygiene Cosmetic Paper Manufacturing Company first promoted its "Clean Pad" in South Korea, it claimed that its sanitary pad was "very popular in the United States and several European countries." [31] Ten years later, in advertising to the Korean market, the Hawgang Trading Company likewise referenced the tampon as a technology already in use for quite some time in Western countries.[32]

As the interview with Ohyeon-daek indicates, mass-produced sanitary pads and tampons entered a context rife with options for South Korean women: these would-be consumers fashioned menstruation pads in various shapes, from different materials. The techniques used by women to fold and to fasten the pads were also highly individual.[33] The pads were referred to by different nouns, depending on the region of Korea and the extent of the taboo around menstruation in that region. Perhaps the most telling of these denominations was, simply, "laundry" (*sŏdap*; 서답); the generic nature of the terms underscores the fact that it was often considered inappropriate and embarrassing to refer directly to the technology in question.[34] Linguists call this common phenomenon "word taboo."[35]

Women experimented with various kinds of fabric to achieve the goal of absorbing menstrual blood. As mentioned, Ohyeon-daek first received from her mother pads made of cotton; later, she used layers of gauze, which she attached to her underwear with rubber bands. Whereas some women used woven cotton to make specially designed pads, other women simply recycled old pieces of cloth or rags. Apparently, wool was less common. Hemp had the benefit of being comparatively easy to wash, though it had lower absorbency than either cotton or wool.[36] Most women designed their own solutions or relied on close relatives for pads.[37] Despite

[31] *Kyunghyang Shinmun* (경향신문) ["Kyunghyang Newspaper"], April 5, 1966.

[32] *Yŏsŏngdonga* (여성동아) ["East Asian Women"], June 1976.

[33] Lee, "Introduction"; cf. transcription from Lee's interview No. 1 (anonymous, born in 1953).

[34] Ibid., p. 8.

[35] Allan, Keith, and Kate Burridge, *Forbidden Words: Taboo and the Censoring of Language*. Cambridge: Cambridge University Press, 2006.

[36] Lee, "Introduction," p. 10.

[37] Lee, "Introduction"; cf. transcription from Lee's interview No. 5 (anonymous, born in 1961).

the taboos around the topic, it is likely that mothers, grandmothers, and older sisters, for example, helped girls cut, sew, fold, and fasten the pads.

In this context, in which mass production and mass consumption did *not* rule the economy, self-made artifacts were paramount. Fittingly, de Grazia refers to the mass-producing economy as the Fordist consumption regime.[38] According to standard academic thinking, Henry Ford—and Fordism—are associated with product standardization and large-scale manufacturing. De Grazia takes this concept further, pointing out that the "Fordist consumption regime" relies on well-established distribution and retailing networks if it is to succeed. When we apply this concept to the mass consumption of disposable sanitary pads and tampons, we realize that mass production is logical only if "the distribution problem," as de Grazia calls it, has been resolved. The Park administration strove to restrict the expansion of department stores and chain stores—and thus the emergence of a Ford-like mass-consumption regime.

The do-it-yourself (DIY) culture around sanitary pads in Korea continued throughout the 1960s. The availability of mass-produced, disposable pads did not appeal to large numbers of women. Occasional, early newspaper advertisements indicate that reusable (non-disposable) feminine-hygiene products were already available during the period of Japanese occupation (1910–1945). Despite this, historian Youngju Lee observes that, well into the 1960s, even wealthy women continued to use home-made menstruation technologies.[39]

To turn a piece of cloth into a useful menstruation napkin requires experience and skill. Through a series of six photographs, Baek documents how one of her interviewees, Bonghwa-daek, uses her experienced hands to transform a rectangular piece of cotton into a thick and stable pad. For those who would like to try it out, Baek provides a series of thirteen sketches which in detail depicts the complicated folding technique. To make sure the sanitary pad stays in position and does not fall out of the underwear, Bonghwa-daek makes two holes in the piece of cloth and leads a rubber band through them.[40]

[38] Grazia, Victoria de, "Changing Consumption Regimes in Europe, 1930–1970: Comparative Perspectives on the Distribution Problem," in: Strasser, Susan, Charles McGovern, and Matthias Judt, eds, *Getting and Spending: European and American Consumer Societies in the Twentieth Century*. Cambridge: Cambridge University Press, 1998: 59–83.

[39] Lee, "Introduction," p. 9.

[40] Baek, "Creation," pp. 55–56.

Gopyeong-daek, another interviewee of Baek's, illustrates an alternative folding technique. Gopyeong-daek uses an almost square piece of fabric and folds it to create an elongated pad. Compared to Bonghwa-daek's fairly thick napkin, Gopyeong's design is much thinner and has a form that reminds us more of today's disposable pantyliners—with the important difference that it is kept in place by a rubber band rather than by an adhesive layer.

Other sources indicate that some women folded their pads in such a way that it was thicker in the front part and thinner at the back.[41] Some women inserted extra cotton or a piece of cloth into a pocket in the middle, to increase absorbency.

Why did the majority of women in Korea decline to adopt disposable sanitary pads in the 1960s and 1970s? The answer relates to the prevailing fashion: at the time, women in Korea preferred skirts and dresses to trousers. Accordingly, women were comfortable wearing traditional underpants that were baggy and loose, rather than form-fitting—and, the use of self-adhesive pads does not work well in baggy underpants. Only when women began to adopt US-style blue jeans, for example, was it necessary to consider wearing form-fitting panties.[42] And only then would it be worth considering the use of Western-style sanitary pads. Traditional underwear was incompatible with disposable pads. So, other technologies were required. Typically, rubber bands or thread was used to fix the sanitary pad to the underpants. Another option was to use buttons to attach the pads to the baggy underwear. Rubber bands and string could also be sewn into the pad to keep it in place. Safety pins were also used. To make the pad more easily attachable, one solution was to sew small hooks or rings into the underpants, allowing for thread or rubber bands to be

[41] Jeon, Gyeong-ok, et al., *Hangug-yeoseong-inmulsa1 - hangug-yeoseong-geunhyeondaesa1. hangaehwagi~1945nyeon* (한국여성인물사1 - 한국여성근현대사1. 한개화기~1945년) ["The History of Korean Female Figures 1: The Modern and Contemporary History of Korean Women 1: The Han Period till ca. 1945"], Seoul: Asian Women's Research Institute, Sookmyung Women's University, 2004, p. 256.

[42] Concerning the growing influence of Western fashion in South Korea in the postwar era, see Kim, Soo-Jin, "Yŏsŏngŭibok-ŭi pyŏnch'ŏn-ŭl t'onghae pon chŏnt'ong-kwa kŭndae-ŭi chendŏjŏngch'i - haebang ihu ~ 1960 nyŏndae ch'oban-ŭl chungsim-ŭro" (여성의복의 변천을 통해 본 전통과 근대의 젠더정치-해방 이후~1960년대 초반을 중심으로) [authorized translation: "Reading the Transformations of Women's Clothes in Postcolonial Korea: Gender Politics of Tradition and Modernity"], *P'eminijŭ myŏ n'gu* (페미니즘연구) ["Issues in Feminism"] 7 (2), 2007: 281–320.

used.[43] A minority of women chose to invest in a commercially available, reusable menstruation belt, to which the self-made pads could be attached. Newspaper advertisements feature such belts—mostly made of a rubber-like material—from the 1920s onward.[44]

Women who used self-made sanitary pads faced the monthly problem of laundering them. Those who lived in rural areas could go to a stream outside the village to do the laundry, as Ohyeon-daek did. Others did their laundry at home, surreptitiously, hanging the pads to dry between other items of clothing, to hide them from view. In 1955, a medical doctor recommended using an iron to disinfect sanitary pads after they had been washed.[45]

Research conducted with women in Taiwan during the mid-1980s is consistent with the research on menstruation technologies carried out with women in South Korea. Nurses and anthropologists in Taiwan found that it was just as important for women there to conceal menstrual blood as it was for women in Korea to do so. Washing sanitary pads was a difficult, burdensome task, especially for those who lacked access to running water. Older women from Taiwan recall how, in the 1930s, women washed their self-made sanitary pads "in ditches and streams." Whereas many of the women in extremely low-income, rural areas appear not to have used any specialized menstruation technologies at all, women of means "liked grass paper because it could be thrown away." The disadvantage was that grass paper, which can be made from hay or grass clippings, was "stiff" and scratchy.[46] The use of grass paper in Taiwan is noteworthy, as it shows that disposability is not a feature exclusive to industrially manufactured products.

Comparisons between feminine-hygiene products in South Korea and those other East Asian countries yield some unexpected differences. For example, consumers in the (then) Republic of China had the option to buy Kotex, Modess, and other disposable sanitary pads as early as the

[43] Jeon, "History," p. 209.

[44] Such ads from the interwar era can be found in the newspaper *Dong-A Ilbo* (동아일보); see Lee, "Introduction," Footnote 5.

[45] Lee, Hwa-hyung, et al., *Han'gukhyŏ ndaeyŏ sŏ ng-ŭi ilsangmunhwa 8. Kajŏ ngwisaeng* (이화형 외, 한국현대여성의 일상문화 8. 가정위생) ["The Daily Culture of Modern Korean Women, 8: Home Hygiene"], Seoul: Kuk'akcharyowŏn, 2005, p. 251.

[46] Furth, Charlotte, and Ch'en Shu-yueh, "Chinese Medicine and the Anthropology of Menstruation in Contemporary Taiwan," *Medical Anthropology Quarterly* 6 (1), 1999: 27–48; here: 41–42.

second half of the 1920s, shortly after these products first appeared on the US market.[47] It would be forty years before these items reached the South Korean market, and another decade would pass before Kotex became available in Taiwan.[48]

In interwar China, when disposable pads were on the market, articles that openly discussed "female hygiene" also appeared in various magazines.[49] According to historian Shing-ting Lin, medical doctors, journalists, and readers in the 1920s broached "bleeding, cramps, sex and sanitary napkins with all the frankness long devoted to digestion and respiration."[50] The Shanghai-based "Ladies' Journal" not only accepted advertisements for Kotex and other disposable sanitary pads, but the publication also allocated editorial space to an increasingly Western discourse on health and hygiene. Obliquely referencing Western knowledge and expertise, promoters of Kotex claimed the disposable pad was "born out of scientific advances and tested by medical professionals."[51] A 1928 Kotex ad juxtaposes these industrially manufactured pads with so-called "unhygienic rags," creating a comparison that apparently resonated with many readers. In letters to the editor, readers often described traditional menstruation technologies as "filthy" and "coarse."[52] Not everyone shared this view, however. In one letter, Zhen San criticized the industrially produced pads, instructing readers on how to make their own version:

> Use sterilised or salicylic acid cotton wool, three or four centimetres long, to press against the part [i.e. the vulva]. Fasten it in the crotch with a cotton belt folded into a T-shape. Fix the device at the waist, using lengths of string at front and rear. [...] When experiencing an excessive loss of blood, change [the pad] two or three times a day. The used pad [is] discarded.[53]

[47] Lin, Shing-ting, "'Scientific' Menstruation: The Popularisation and Commodification of Female Hygiene in Republican China, 1910s–1930s," *Gender & History* 25 (2), 2013: 294–316.

[48] Wang, Hsiu-Yun, "Postcolonial Knowledge from Empires: The Beginnings of Menstrual Education in Taiwan, 1950s–1980s," *East Asian Science, Technology and Society* 11 (4), 2017: 519–540, here: 533.

[49] Lin, "'Scientific.'"

[50] Ibid., p. 294.

[51] Ibid.

[52] Ibid., pp. 294, 307.

[53] Translation of a passage from an article in "Ladies' Journal" (1928) by Zhen San; quoted from Lin, "'Scientific,'" p. 307.

Interestingly, Zhen's solution represents a middle-ground between recyclability and disposability. The cotton belt, which probably resembled Gopyeong-daek's design, was washed and reused repeatedly, whereas the small piece of absorbent cotton wool was thrown away.

Mass-Produced Pads: A Hard Sell

As we have seen, advertisements for the first disposable sanitary pads in South Korea referred explicitly to their Western origins. Decision-makers in marketing departments appear to have been convinced that the association with the United States and Europe would persuade potential users—especially young, urban women—to abandon self-made pads in favor of "modern" products. The time seemed right: people's discretionary income was surging, and many young men and women were willing to spend more on products considered luxuries by their parents. Stoked in part by television and other mass media, a Western-inspired version of consumer culture was emerging.[54] Increasingly, more women in Korea entered the workforce; in the 1960s and 1970s, their numbers more than doubled.[55] To enable women to withstand many consecutive hours of working on production lines and in offices, disposable pads proved their convenience. In most cases, it was more difficult to hide or wash sanitary pads in the workplace than it was to do so at home.[56]

As mentioned, the Clean Pad brand was one of the first on the South Korean market. "Anemone" was another, manufactured and marketed by the Seoul Paper Manufacturing Corporation. A 1970 advertisement published in a Korean women's magazine does not mention any connection between the product and the United States or Europe; rather, the ad highlights Anemone's comfort and convenience, claiming, "It is easy to use";

[54] Cho, Haejoang, "Living with Conflicting Subjectivities: Mother, Motherly Wife, and Sexy Woman in the Transition from Colonial-modern to Postmodern Korea," in: Kendall, Laurel, ed., *Under Construction: The Gendering of Modernity, Class, and Consumption in the Republic of Korea*. Honolulu, HI: University of Hawai'i Press, 2002: 165–195.

[55] Ahn, Jae-Hee, "Analysis of Changes in Female Education in Korea from an Education-Labor Market Perspective," *Asian Women* 27 (2), 2011: 113–139; Moon, Seungsook, *Militarized Modernity and Gendered Citizenship in South Korea*. Durham, NC: Duke University Press, 2005, Ch. 3.

[56] Substantial parts of this section are based on Lee, "Introduction," pp. 13–21.

"soft"; "small"; and "you can throw it away in the WC." The ad also touts Anemone as industrially produced—not made by hand.[57]

Although the Anemone ad does not refer explicitly to the West, it does reflect trends found in North America and Europe at the time. When the philosopher Herbert Marcuse in 1964 published his groundbreaking book *One-dimensional Man*, he chastised what he calls (in the subtitle) *Advanced Industrial Society* for preventing us from being active, politically aware citizens, and turning us all into passive consumers. Claiming that the Western world had given in to the ideals of consumerism and convenience, Marcuse refers to both the constant stream of trivial entertainment—symbolized by the emerging medium of television—and to the "capacity" of "industrial civilization … to increase and spread comforts, to turn waste into need."[58] The consumption of disposable products ("waste"), likely belongs to the comforts Marcuse and later thinkers had in mind.[59] In the context of menstruation technologies, the relevant question is whether women in South Korea were swayed by the promises about disposable pads—and later, tampons—made in advertisements.

In fact, most women in Korea were not immediately convinced of the advantages of the Clean Pad, Anemone, nor other disposable sanitary pads. The companies' arguments—that their products were "healthy," "easy," and "safe" to use—failed to persuade.[60] Youngju Lee presents research from the late 1960s indicating that, calculated per-capita, a Korean woman in her reproductive years used only 2.5 disposable pads annually. This may suggest that less than one percent of women had adopted this new, disposable-pad technology. Not even Kotex—launched in Korea fifty years after its introduction in the United States—was an immediate market success. Yuhan-Kimberly was a joint venture between Kimberly-Clark (US) and the Yuhan Yangheang Corporation (South Korea), which produced pharmaceutical and chemical products. In 1970, in a factory near Seoul (Anyang-si), Yuhan-Kimberly manufactured both Kotex sanitary pads and Kleenex tissues. To comply with the Park administration's economic policy, the company was obliged to export a substantial proportion of its products. In a corporate history of the Yuhan-Kimberly

[57] *Yŏsŏngdonga*, Jan. 1970.

[58] Marcuse, Herbert, *One-dimensional Man: Studies in the Ideology of Advanced Industrial Society*. London: Routledge & Kegan Paul, 1964, p. 9.

[59] Cf., e.g., Shove, Elizabeth, *Comfort, Cleanliness and Convenience: The Social Organization of Normality*. Oxford: Berg, 2003.

[60] *Kyunghyang Shinmun*, April 5, 1966.

company, an employee remembers how difficult it was to market the mass-produced sanitary pad in South Korea. It took considerable effort to convince pharmacists to carry the new product. Similarly, when a company representative attempted to hand out free samples of the sanitary pad to women on a bus, the driver forced the salesperson off the vehicle.[61]

To generate sales, Yuhan-Kimberly from the start attempted to influence women by referencing the international success of Kotex. One of the company's first advertisements, from 1971, shows a young Korean woman with a globe of the world in the background; the accompanying text informs would-be consumers that women in 129 countries were already using Kotex.[62] Soon after, the company produced an ad that posed the seemingly rhetorical question of why women in Korea persisted in using their "traditional" sanitary pads. The text of a 1973 advertisement questions why not all women in Korea have abandoned the "uncomfortable" task of washing bloodstained fabric every month. Apparently, the company was betting on the idea that laundering self-made sanitary pads was onerous, as well as too intimate a task to be delegated to others—so they would soon realize that Kotex pads were far more "convenient and safe." Featuring a photograph of a women styled as a housewife, the ad implies that only women in traditional roles would insist on using self-made sanitary solutions, which required washing.[63]

On a conceptual level, then, the Yuhan-Kimberly marketing department reflected Marcuse's values of an advanced industrial society. The question is, why were women in Korea largely relatively indifferent to the promise of "comfort" and "convenience" as it related to industrially produced menstruation products?

How did financial constraints play a role in women's consumption (or non-consumption) of disposable sanitary pads in Korea? One theory is that, to economize, women bought disposable pads to use only for outdoor activities; on holidays; and while traveling. In a 1970 article in the

[61] Moon, Kook-Hyun et al., *Segye-ka paeunŭn han'guk kiŏp-ŭi hŭimang YuhanKimberly* (세계가 배우는 한국 기업의 희망 유한킴벌리) ["A Role Model for Korean companies: Yuhan-Kimberly that the World is Following"]. Seoul: Hans Media, 2005, p. 91.

[62] *Yŏsŏngdonga*, May 1971.

[63] Ibid., Oct.-Dec. 1973. Cf. the analysis in Roh, Jieun, "1970 nyŏndae~1990 nyŏndae saengnidae kwanggo tamnon-kwa yŏsŏng (1970년대~1990년대 생리대 광고 담론과 여성)" [authorized translation: "1970s~1990s Menstrual Product Advertisements and Gender in Korea"], *Yŏsŏng-kwa yŏksa* (여성과 역사) ["Women and History"] 21, 2014: 219–49; here: 233–234.

magazine *Yŏsŏngdonga* ("East Asian Women"), Kang Ji-yong, a medical doctor at the Ewha Womans [*sic*] University, recommends using traditional, self-made sanitary pads at home and at work, and switching to disposable pads only when standard routines cannot be followed. Given her belief that "conventional" pads are safer and more efficient than industrially produced ones, Dr. Kang—a teacher at the College of Medicine—suggests refraining from using disposable pads while asleep.[64] This recommendation may relate to the fact that the disposable pads available at the time lacked an adhesive strip to keep it in place.[65]

Further sources corroborate the observation that, for a long interval, disposable sanitary pads failed to render obsolete the traditional self-made options; the new and old technologies existed side by side, often complementing each other. In 1968, the editors of *Yŏsŏngdonga* presented readers with a service-oriented article on how to wash sanitary napkins "easily."[66] One of the magazine's readers submitted a letter in which she shares her advice: in the absence of access to boiling water, one can wash the sanitary napkins in lukewarm water to which sodium carbonate (washing soda) has been added. Notably, three years later, an Ewha Womans University textbook on "sex education," written for middle- and high-school pupils, refrains from recommending the use of either self-made or store-bought, disposable pads. While they deem disposable pads "convenient," the authors go on to encourage every girl to decide for herself which kind of pad to use.[67] (Not considered a serious option, tampons are mentioned only in passing.) One of Lee's interviewees, whose menstrual cycle first began in 1980, explained that she used disposable pads only when she was away from home; otherwise, she reverted to self-made sanitary pads. This was primarily for financial reasons.[68]

Moral concerns made it difficult for manufacturers of disposable pads to reach large consumer groups. For example, a 1973 article in *Women Dong-A* sharply criticizes Yuhan-Kimberly's television commercials for Kotex. The article's author, a self-described housewife named Kim Sŏng-suk, reports finding it highly embarrassing to be compelled to watch such

[64] *Yŏsŏngdonga*, June 1970.

[65] Lee, "Introduction," interviewee No. 12 (anonymous, born in 1967).

[66] *Yŏsŏngdonga*, June 1968.

[67] *Chung· kogyosaeng-ŭl wihan sŏn'gyoyuk – kyehoek-kwa siltche* (중·고교생을 위한 성교육-계획과 실제) ["Sex Education for Middle and High-school Pupils: Plan and Practice"]. Seoul: Ewha Womans University, 1971, p. 74.

[68] Lee, "Introduction," interviewee No. 7 (anonymous, born in 1965).

advertisements together with male and female relatives. She notes that the commercial caused discomfort to everyone in her company.[69] Criticism of the commercials continued, and in 1980, television commercials for menstruation products were, in fact, banned in South Korea.

Despite these public concerns, textbook authors gradually accepted disposable sanitary pads as a standard technology. A nicely illustrated book, published in 1984 and aimed at children in fifth and sixth grade, includes a romanticized drawing of an industrially produced pad. Focusing on cleanliness and hygiene, the author advises girls to change pads frequently and to make sure to keep the genital area clean. The author also states that, to reduce the risk of bacteria entering the vagina, menstrual pads should be put in place from front to back. Further advice: hard work and swimming should be avoided during the girl's period.[70]

Freedom to Consume

Like Baek Min Jeong's master's thesis, Youngju Lee's thesis is based on interviews with women in South Korea.[71] Notably, one of Lee's thirteen interviewees, who is quoted anonymously, explains that she used recyclable cloth pads well into the 1990s. This interviewee was born in 1953 and got her period for the first time at age seventeen. After working in the fields and in the home for roughly two decades, at the age of forty she became a factory worker. It was only after one of her coworkers had showed her the advantages of disposable pads that she decided to adopt this technology. In the interview, she states that, had she continued to stay at home, she would not have turned to disposable pads. At the workplace, the disposable pads were more convenient, though she found cloth pads to be softer and smoother.[72]

In general, women in Korea began to accept disposable sanitary pads in the second half of the 1970s. One of Lee's interviewees states that, in the small town where she grew up, disposable sanitary pads were still unavailable throughout the early 1970s. In 1975, newspapers continued to associate disposable pads with urban areas, and cloth pads with rural

[69] *Yŏsŏngdonga*, Sept. 1973.

[70] Lee, Won-gu, *Sojung-han uri-dŭl - 5 · 6 hangnyŏn* (소중한 우리들 – 5 · 6 학년) ["Precious Us: 5th and 6th Grade"]. Seoul: Tongjinmunhwasa, 1984, pp. 81–81.

[71] This section is based largely on Lee, "Introduction," pp. 29–43, including several of Lee's interviews.

[72] Ibid., interviewee No. 1 (anonymous, born in 1953).

areas—although disposables by that time had begun to make inroads even in the countryside.[73] Hailing a new "openness" and the "fresh air" in Korean villages, a journalist in 1977 observes that disposable sanitary pads had become a bestseller at local markets. In an interview, a woman pharmacist from Hwaseong County reports that rural women would walk for hours to the nearest small town to buy disposable pads.[74]

Apparently, this new "openness" was widespread in South Korea. Another of Lee's anonymous interviewees, who was born in 1961 in the southeastern part of the country but had grown up in Seoul, relates that she had no problem talking to her mother about her period and various personal-hygiene techniques.[75] Meanwhile, public criticism continued, with some people citing the negative impact of TV commercials—for sanitary pads and contraceptives—on children and youth.[76] This criticism did not staunch the trend toward more candid treatment of these formerly taboo topics. Commercial interests were only partly responsible for the changes; the Park administration also had a hand in this shift. The textbook mentioned above, *Sex Education for Middle- and High-School Pupils: Plan and Practice*, had been supported by the Ministry of Culture and Education; the book reproduces many of the arguments found in the government's so-called Family Planning Program. Supported by a number of international organizations, this program aimed to reduce the number of births in South Korea. To this end, it was deemed essential that young citizens receive the relevant information about the biology of reproduction and the technology of contraception. Knowledge about menstruation and personal-hygiene technologies was a part of this package.[77]

Throughout the 1970s, manufacturers of disposable personal-hygiene products, including tampons, intensified their marketing efforts. New companies entered the market. Yungjin Pharm introduced a pad named Sophia, and the Ilyang Pharmaceutical Industry Corporation, in collaboration with a Japanese company, developed a product called Charming Chanel. The Hawgang Trading Company was the first manufacturer of tampons in South Korea. Its Mon Clean was soon followed by the Amore tampon, manufactured by the Pacific Chemical Company (now the Amore

[73] Ibid., interviewee No. 2 (anonymous, born in 1970); *Dong-A Ilbo*, 17 Oct. 1975.

[74] *Kyunghyang Shinmun*, Aug. 11, 1977.

[75] Lee, "Introduction," interviewee No. 5 (anonymous, born in 1961).

[76] *Dong-A Ilbo*, April 27, 1979.

[77] DiMoia, John P., *Reconstructing Bodies: Biomedicine, Health, and Nation Building in South Korea since 1945*. Stanford, CA: Stanford University Press, 2013, Ch. 4.

Pacific Corporation) and the Tempo tampon, made by the Dong-A Pharmaceutical Company.

Intense competition followed. To raise awareness of their products among women, company representatives canvassed workplaces and distributed free samples.[78] To cultivate future customers, companies also offered schools across the country special classes in sex education, an offer many teachers happily accepted.[79] Advertisements in magazines and newspapers, as well as television commercials (up to 1980), reproduced the Western image of disposable pads. Yuhan-Kimberly picked up this trope—the association of sanitary pads and tampons with a Western lifestyle—in several advertisements for its disposable pad named New Freedom. In one ad, to reinforce the association with the United States, the name of the product is written in English—in capital letters, no less. In another ad, a well-known Korean actress appears in a stars-and-stripes sleeveless shirt.[80] Engaging another famous actress in Korea to promote Sophia, the Yungjin Company explicitly labeled its product "European." To highlight the sanitary pad's identity as European, the featured actress's style was unmistakably inspired by Western fashion: brown, curly hair; dark eyeshadow; and a turtleneck.[81]

The Yuhan-Kimberly and Yungjin companies appear to have been particularly successful in achieving brand recognition in the market. One of Lee's interviewees recalls that she and her friends seldom used the Korean word for menstruation or sanitary pad (*saengnidae*; 생리대). Instead, they simply referred, respectively, to "Sophia" and "Freedom."[82]

The brand name New Freedom was not meant to be understood in the political context of Western democratic rights; after all, the Park administration was a military regime that monitored closely people's lives as well as the country's companies. Rather, manufacturers aimed to highlight the freedom to enjoy newfound leisure activities. This marketing strategy squared well with the growing importance of domestic tourism and

[78] Lee, "Introduction," interviewee No. 12 (anonymous, born in 1967)

[79] Cf. oral evidence in Roh, Jieun, "Wŏlgyŏng kyŏnghŏm-kwa munhawjŭk kŭmgi" (월경 경험과 문화적 금기) [authorized translation: "A Study of the Experiences of Menstruation and the Cultural Taboos"], *MA Thesis*, Seoul: Ewha Womans University, 1995, p. 34, as well as in Lee, "Introduction," interviewee No. 3 (anonymous, born in 1959).

[80] *Kyunghyang Shinmun*, March 10, 1975; *Yŏsŏngdonga*, Oct. 1975.

[81] *Yŏsŏngdonga*, March 1976.

[82] Lee, "Introduction," interviewee No. 3 (anonymous, born in 1959).

weekend excursions in the 1970s.[83] One Yuhan-Kimberly ad states that New Freedom enables a "Gentle Life"; another ad makes the grandiose claim that the Maxi version of the pad will "MAXIMIZE new freedom."[84] Interestingly, both ads make partial use of the English language—and Latin alphabet—to strengthen the association with the West. Similarly, an ad for Kotex pads suggests the product enables Korean women to enjoy "Hip Fashion"—written in English.[85] In some ads, Yuhan-Kimberly emphasizes the connection with the cultures of the United States and Europe in the 1970s by using Flower Power–inspired drawings commonly associated with Western pop music and alternative, hippie lifestyles.[86] To ensure freedom of movement, New Freedom was the first disposable pad with an adhesive strip. This design was meant to secure the pad in the underwear, obviating the need for hooks, buttons, or rubber bands. The new pad design implied that women would be able to go to the movies and to concerts, to play sports, and more. Thus, disposable pads were not simply Western; they gave women the opportunity to carry out their daily activities, even during their periods. With the adhesive-strip pad, women would be relieved of worry about having accidents with their sanitary pads, both at work and during their leisure time. The self-adhesive pad dovetailed nicely with South Korea's US-style consumer-society-in-the-making.

In a history of feminine-hygiene-product advertising, Feminist Studies scholar Jieun Roh relates disposable sanitary pads to changes in leisure activities and fashion. Citing further examples in the magazine *Yŏsŏngdonga*, Roh references manufacturers' claims that, with their products, women could enjoy freedom of travel—throughout the month.[87] The idea was that women were free of leaky cloth pads that could prevent them from taking vacations and other trips, as well as playing a range of sports. For example, Yungjin Pharm's ads for the Sophia pad feature images of women playing baseball; fencing; and riding horseback.[88] Advertisements by various other manufacturers associate disposable pads with bicycling and sailing, roller-skating and skiing, for example.[89]

[83] Song, "Politics."
[84] *Yŏ sŏ ngdonga*, Feb. 1976.
[85] *Yŏ haksaeng* (여학생) ["Female Students"], Nov. 1980.
[86] Ibid., Aug. 1978; May 1979.
[87] Roh, "Menstrual Product," pp. 238–240.
[88] *Yŏ sŏ ngdonga*, April–May 1976; *Yŏ haksaeng*, Nov. 1979.
[89] *Yŏ haksaeng*, Nov. 1977, April 1978.

In the mid-1970s, when tampons were introduced to the Korean market, advertising for the product featured sports and other leisure activities even more prominently than ever; Roh demonstrates that this continued well into the 1990s.[90] For example, ads sponsored by the Dong-A Pharmaceutical Company purport that tampon use allows girls and women to play tennis and to engage in mountain climbing.[91] Featuring six photos of young, White women from Western European as well as Anglo-Saxon countries, one Dong-A ad suggests that tampon use will enable Korean women to enjoy activities associated with a Western lifestyle; women can now go swimming and skating during their periods. In this ad as in others, the company uses the English slogan "TEMPO DAY" to highlight how tampons give women the freedom to plan their daily activities as they like.[92] Manufacturers targeted girls and young women in particular, urging them to seize the day before traditional, adult responsibilities curtailed their freedom.[93]

Tampon ads repeated the same promises as the advertising for disposable pads: comfortable and convenient to use, tampons gave women a new kind of freedom. In addition, tampons were reputed to be hygienic and leak-proof, as well as to provide protection against unpleasant odors.[94] Tampons were the perfect technology for women who wanted to live an "active" life.[95] Supposedly, this was a life already attained by tampon-using women all over the world. To support this marketing strategy, the magazine *Yŏhaksaeng* ("Female Students") ran ads accompanied by illustrated articles on how to use tampons.[96]

An advertisement for Amore tampons also exemplifies the tampons-equal-freedom narrative. A color photo shows a young woman on a bicycle, coasting through a puddle, her legs in the air to avoid getting splashed. The ad copy tells us it is early morning, the dew is still in the fields, and it promises to be a beautiful day. The reason for the young woman's good mood is also addressed: she is using a tampon, so she can enjoy the beauty of nature via bicycle—without having to worry about her period. The ad also boasts two unusual features of the Amore tampon. First, it does not

[90] Roh, "Menstrual Product," Section III.2.
[91] *Yŏ sŏ ngdonga*, June 1976, July 1977.
[92] Ibid., Feb. 1979.
[93] *Yŏ haksaeng*, March 1979.
[94] *Yŏ sŏ ngdonga*, Dec. 1978, June 1979, Oct. 1981.
[95] *Yŏ haksaeng*, May 1975.
[96] Ibid., Oct. 1977.

require an applicator. Second, it comes from West Germany (manufactured by Dr. Carl Hahn Ltd., known in Germany under the brand name o.b., *ohne Binde*, i.e., "without pad") rather than the United States.[97]

Although the oral contraceptive pill ("the pill") was, of course, intended primarily to prevent pregnancy, it was also marketed as a technology to control menstruation, thereby affording users more freedom. The Chongkundang Pharmaceutical Corporation claimed that its oral contraceptive pill Norinil would help women to "normalize" menstruation.[98] Presumably, this referred to the idea that, at least for some women, taking the pill would ensure their menstrual cycle occurred every twenty-eight days. The idea of normalizing menstruation may also refer to the fact that, for most women, the pill reduces bleeding considerably.

Resisting "The American Empire"

Undoubtedly, the United States exerted considerable influence on consumer culture in South Korea. Magazines published ads for several US (and other Western) products, including Coca-Cola and Fanta.[99] To help sell their refrigerators, General Electric's ads sported a modified American flag.[100] The US actresses Katharine Ross, Deborah Raffin, and Sydne Rome, appeared in Yuhan-Kimberly advertisements for disposable sanitary pads.[101]

Indeed, manufacturers of sanitary pads and tampons used as a key selling point the assertion that women in the United States, in European countries, and elsewhere had already adopted the products. Although it was repeated frequently in advertisements and television commercials, the message slowly lost its persuasive power; marketers began to realize that target customers might well be more susceptible to other campaign messages. While many advertisements and commercials for the Korean market reproduced US media campaigns based on the benefits of freedom, safety, and comfort, some marketing strategies were based on the needs of the Korean market.[102] For example, a 1979 advertisement proclaimed that the

[97] Ibid., Jan. 1979.
[98] *Yŏsŏ ngdonga*, Oct. 1978.
[99] Ibid., May 1970; Aug. 1973.
[100] Ibid., Jan. 1980.
[101] *Yŏhaksaeng*, March & June 1975, July 1976.
[102] Concerning advertisements in US magazines, see Merskin, Debra, "Adolescence, Advertising, and the Ideology of Menstruation," *Sex Roles* 40 (11–12), 1999: 941–957.

Tempo tampon had been redesigned to better fit women in Korea. In particular, the company shortened the tampon and softened the applicator, claiming that these changes would make the tampon easier to insert.[103] By touting the tampon as customized to Korean consumers, marketing people apparently tried to make tampons look and feel less "foreign," as it were. This strategy did not succeed, however.[104] Statistics show that the tampon market began to take off only around 2010—more than thirty years after the product was first introduced in South Korea.[105]

The subtext of a 1978 Tempo advertisement indicates another reason for the slow acceptance of tampons by women in Korea. The ad, sponsored by Dong-A Pharmaceutical, mentions that "even unmarried women" can use Tempo.[106] This statement addresses the fear, believed to exist among young women, that tampon use would damage the hymen. As letters to the editor of the magazine "Female Students" show, this fear was indeed real. In 1985, for example, a "middle-school student" claims to have heard from a friend that "wrong usage" of a tampon might hurt the hymen. In a worried tone, she asks if a damaged hymen would make it impossible for her "to bear a baby" later in life.[107] It is impossible to know the extent to which this anxiety was a motivating factor in young women rejecting the tampon. Jieun Roh claims that the rejection of tampons by unmarried women reflects a paternalistic ideology. According to this thinking, a bridegroom would expect his bride to be a virgin on their wedding night, and an intact hymen supposedly represented proof of virginity. Although this ideological interpretation may well have played a role, Youngju Lee cites several other reasons why many women in Korea—regardless of age and marital status—chose not to adopt the use of tampons. As we have seen from their marketing campaigns, tampon manufacturers suggested tampons would enable women to take part in sports during their period. To the interviewees, however, this "new freedom" to take part in sports every day of the month infringed on their freedom to *refrain* from intensive physical activity and labor when they were menstruating. For example, some of Lee's interviewees recall being

[103] *Yŏhaksaeng*, Oct. 1980.

[104] Concerning the concept of "domestication" in technology studies, see Lie, Merete, and Knut H. Sørensen, eds, *Making Technology Our Own? Domesticating Technology into Everyday Life*. Oslo: Scandinavian University Press, 1996.

[105] Roh, "Menstrual Product."

[106] *Yŏsŏngdonga*, Oct. 1978; cf. Roh, "Menstrual Product," p. 83.

[107] *Yŏhaksaeng*, June 1985, p. 328.

relieved at the prospect of ***not*** having to attend gym classes during the peak of their periods. By rejecting tampons, Lee suggests, women affirmed the specificities of their own bodies and their gender identity, thus rejecting the idea that women need to perform in the same way as men. Women in the 1980s began to embrace the fact that their bodies were distinct and special. By choosing pads over tampons, women in Korea distinguished themselves from what they viewed as Western: rationally performing women's bodies.[108]

Despite women's newfound appreciation for their own bodies in the 1980s—according to Lee's interpretation—manufacturers continued to insist on the rational advantages of sanitary pads in the ensuing decades. In step with the increasing numbers of women in non-manual-labor jobs, advertisements targeted career women rather than girls and athletic women. The ad campaign's reasoning was: if a woman stayed home during her period, she would have no chance of winning in the fierce competition with her male colleagues. The new disposable sanitary pads and tampons might not erase physical pain and fatigue, but they were promoted as making users feel confident of getting through the entire workday—without awkwardness or embarrassment.[109]

A comprehensive survey organized by the World Health Organization (WHO) in the mid-1970s supports the argument that women in South Korea followed a different path from their counterparts in the West. Whereas in the UK, almost half of women of reproductive age used tampons, in Korea, the proportion of tampon users was so low that it failed to show up in the WHO statistics; the US figures were comparable to those in Britain.[110] A plausible explanation for the reported low tampon use in Korea is that only three percent of the WHO interviewees were under twenty-four years of age. Although more than half of the 500 Korean women who took part in the survey lived in urban areas, seventy-one percent of the entire sample used homemade sanitary pads; only twenty-six percent of the women interviewed used "store-bought" pads.[111]

108 Cf. Roh, "Menstrual Product," and Lee, "Introduction."

109 Roh, "Menstrual Product," Section III.3.

110 Seok-jin Jang, "Urinara yŏsŏng saengniyongp'um chejoŏp-ŭi maak'et'ing-e kwanhan yŏn'gu" (우리나라 女性 生理用品 製造業의 마아케팅에 관한 硏究) ["A Study on the Marketing Female Menstrual Products in Korea"]. *Master's Thesis*, Korea University School of Management, 1981, p. 27.

111 Snowden, Robert, and Barbara Christian, eds., *Patterns and Perceptions of Menstruation: A World Health Organization International Collaborative Study*. London: Croom Helm, 1983, pp. 11, 53, 249.

In summary, considering the behavioral trends of women in South Korea, the United States was *not* viewed as an "irresistible empire." US-designed personal-hygiene products such as disposable sanitary pads and tampons became available to Korean consumers only in the 1960s (sanitary pads) and 1970s (tampons)—an era in which a Western-inspired consumer culture began to emerge. To establish their products on the Korean market, some US companies entered joint ventures with Korean partners, establishing domestic-production facilities. Although these efforts were modestly successful, the market for disposable products did not reach the levels Western companies were accustomed to achieving in North America and Europe.

Finally, we must answer the question of why women in South Korea "resisted"—or at least did not wholeheartedly accept—US consumer culture when it came to grappling with menstruation. The answer, which complements Lee's conclusion, can be extrapolated from yet more WHO data: four out of five of the interviewees in Korea interpreted menstruation as a sign of femininity, whereas only a small minority regarded menstruation as "dirty" or as a "sickness." In contrast to Baek's observations about the taboo nature of menstruation as a topic, two out of three interviewees claimed to discuss issues around menstruation with their female friends.[112] Thus, consistent with Lee's interpretation, the majority of women in Korea regarded menstruation as a normal, natural event, which they accepted and affirmed. For these interviewees, using a convenient technological fix, such as a tampon, was tantamount to negating their femininity by "hiding" their periods. To embrace the identity of a "Korean woman" was apparently more desirable than to become an "American lady." In a 2004 journal article, feminist scholar Soojung Kim expresses a similar idea. Kim argues in favor of returning to the use of washable sanitary pads—a practice that, by then, had become what she calls an "alternative technology."[113]

The story of self-made, reusable sanitary pads in Korea runs contrary to the evolutionary theory of technological development; products that were once discarded in the name of convenience or rationality may well return—regardless of whether we call them "old," as David Edgerton once did, or "alternative," as Kim does.[114]

[112] Snowden and Christian, *Patterns*, p. 58.

[113] Kim, Soojung, "Tarŭn sesang : Kaejim, taean saengnidae iyagi" (다른 세상: 개짐, 대안 생리대 이야기) ["Alternative World: A Story of an Alternative Sanitary Napkin"], *Yŏ/songi-ron* (여/성이론) ["Journal of Feminist and Sexual Theories"] 10 (7), 2004: 209–215.

[114] Edgerton, David, *The Shock of the Old: Technology and Global History since 1900*. London: Profile, 2006.

CHAPTER 9

Doing It Yourself in Central Asia: Uzbeks Build Adobe Houses

> Everything was in bad condition; well, why was everything derelict? It had once been very well-built, with a stable foundation. But then it began to deteriorate, to sag and buckle. This section of the house had been very well constructed. Every generation slowly, slowly built and rebuilt it, but because it was always used, it really began to deteriorate. In principle, every generation—or, at least, every other generation—probably would have had to rebuild this part.[1]

In this 2018 interview, Dilya, a woman in her mid-fifties, reflects on the history of the house she co-owns with her older sister, Rano. Along with their parents and grandparents, Dilya and Rano grew up in Samarkand, Uzbekistan. Over the course of the interview, Dilya recalls what their home looked like thirty years earlier, remarking that currently, the house

[1] Unless stated otherwise, the interviews cited in this chapter were carried out in 2018 as part of the research project for which I was the Principal Investigator: "A Global History of Technology, 1850–2000" (ERC AdG 742631). Mariya Petrova and Jonas van der Straeten conducted the interviews in Russian. The interviewees consented to their information being published anonymously by the project researchers; their names have been changed accordingly. Petrova and van der Straeten conducted the interview with Dilya and Rano excerpted above in Samarkand, Uzbekistan, on September 17, 2018. All quotes represent English translations rendered from transcripts of the original interviews. I am responsible for this translation and all other Russian-to-English translations of primary—and secondary—source material in this chapter. Throughout the writing of this chapter, Mariya Petrova's knowledgeable and critical comments proved invaluable.

M. Hård, *Microhistories of Technology*, Palgrave Studies in the History of Science and Technology,
https://doi.org/10.1007/978-3-031-22813-1_9

bears no resemblance to its former state. Today, the premises are in excellent condition, with no part of the building sagging or buckling, no deterioration in sight. The brick walls, floors, and window frames have been renovated; old, carved columns have been once again exposed; the rooms are nicely furnished; the courtyard garden is filled with vines, small trees, and flowers. Readers of *House & Garden*—or practically any other interior-design magazine—would be charmed by the place.

Dilya and Rano's house is situated in the Old Town district of Samarkand; the property has been in the family since the late nineteenth century. Their great-grandfather was a wealthy Tajik merchant who bought the centrally located plot of land and commissioned a house to be built on the premises. After the turn of the century, he added extensions. Dilya and Rano's grandfather inherited the older parts of the house, and his brother's family moved into the newer section. Grandpa was a lawyer who paid little attention to home repair and maintenance. If anything, the girls' parents were even less interested than their grandfather was in investing time and money in the upkeep of the family home: "Daddy did not build anymore […] it was in critical condition and began to fall apart like that." The only maintenance they did was to whitewash the brick walls occasionally and to paint some of the outside walls and parts of the interiors: "We just used light blue. In Soviet times, this was fashionable—in Russian villages, as well."[2]

Another interviewee, Tatiana, also suggests that renovations were mostly cosmetic—prior to at least 1991, the year when the Soviet Union (Union of Soviet Socialist Republics, or USSR) broke up, and Uzbekistan became an independent country.[3] To a large extent, it was lack of money that accounted for the lack of maintenance. Dilya and Rano's father earned a low salary, and there were few, if any, craftspeople among the family's closest relatives. When urgent repairs needed to be made, family members had to tackle the problems on their own. Or, as Dilya put it, "We did it ourselves" (*delali sami*).

"We did it ourselves"—делали сами in Cyrillic letters—is the reprise of many Central Asian interviewees for this chapter. Here, I highlight the ways in which Uzbeks and other inhabitants in the Central Asian area of

[2] The quotations in this paragraph are taken from the interview mentioned above.

[3] Quotation from an interview carried out by Mariya Petrova and Jonas van der Straeten on September 26, 2018, with Tatiana (name changed). The following quotations in this paragraph are taken from the interview mentioned in Footnote 1.

the Soviet Union approached home renovation; I document their best efforts at home repair. Interviews conducted with ordinary people reveal a high degree of self-reliance. After all, not only were financial resources often limited, but construction materials were at times difficult to source, as were skilled craftspeople. Nonetheless, residents usually did not passively wait for someone else—or for the State—to fix the leaky roof or cracked façade: they did it themselves.

As a concept, "Do-It-Yourself," abbreviated as DIY, is known to many readers. Historians of technology have investigated the DIY phenomenon in domains like home improvement, hobbyism, and computer design.[4] In all societies where people have not been able to find—or to afford—professional assistance, they had little choice but to solve technical problems independently. Although the phenomenon is as old as humanity, historian Steven M. Gelber argues that "by the end of the 1950s the very term 'do-it-yourself' would become part of the definition of suburban husbanding," especially in the United States.[5] According to Gelber's interpretation, especially married male homeowners took pride in carrying out minor repairs and undertaking renovation work in their own houses. In the US suburbs, DIY was an individual practice. Residents certainly discussed various solutions with their neighbors, and they informed themselves by consulting manuals and hobby magazines such as *Popular Mechanics*. By and large, however, these men carried out the work themselves, in their own homes. Hardware stores and tool manufacturers happily supported their endeavors.

Do-it-yourself activities have in no way been limited to US suburbs, of course. In recent years, DIY has developed into a broad user movement, and today people get together in repair cafes, exchange information on social media, and publish their experiences on YouTube.[6] Retired professionals show amateurs how to solve technical problems of home repair, from fixing bicycles to rewiring toasters. To underline the cooperative

[4] Maines, Rachel P., *Hedonizing Technologies: Paths to Pleasure in Hobbies and Leisure*. Baltimore, MD: Johns Hopkins University Press, 2009; Tinn, Honghong, "From DIY Computers to Illegal Copies: The Controversy over Tinkering with Microcomputers in Taiwan, 1980–1984," *IEEE Annals of the History of Computing* 33 (2), 2011: 75–88.

[5] Gelber, Steven M., "Do-It-Yourself: Constructing, Repairing and Maintaining Domestic Masculinity," *American Quarterly* 49 (1), 1997: 66–112; here: 67.

[6] Concerning the concept of the "user movement," see Oldenziel, Ruth, and Mikael Hård, *Consumers, Tinkerers, Rebels: The People Who Shaped Europe*. Houndmills, Basingstoke: Palgrave Macmillan, 2013.

character of this emerging movement, activists even launched the concept of "Do-It-Together" as a "post-capitalist" alternative to the individualistic, consumer-oriented DIY trend.[7]

Historians who research socialist countries have tended to analyze the practice of DIY as a necessary component of societies governed by economic and material scarcity.[8] Given that it reduces DIY to an economic activity, this form of analysis is highly limited. As Dilya's representative narrative indicates, ordinary people in Central Asia had multiple reasons for fixing, extending, and renovating their homes. Just like the archetypical US suburban husband, they carried out DIY work on their homes for the sake of pride and for fun, for social recognition and for personal satisfaction, for a sense of communality and solidarity. Historians and anthropologists have described similar activities and processes for people in the Russian part of the USSR as well as those in post-Soviet Russia.[9] For example, from 1966 onward, hobbyists across the USSR devoured the popular magazine *Modelist-Konstruktor*.[10]

Reference material from Central Asia shows that communal activities are well established in local culture. When Dilya said, "We did it ourselves," she meant that family members carried out the necessary home maintenance. As we will see, however, Uzbek, Tajik, and other people in the region had, in fact, for centuries collaborated with close and distant neighbors on collective tasks. Communal work represents a well-practiced custom—before, during, and after the Soviet interlude. Unlike repair cafes, this tradition has a long history and remains an essential element of Central Asian cultures. Dilya's phrase references the collective and communal nature of such work—rather than the more individualistic work

[7] Baier, Andrea, et al., "Die Welt reparieren: Eine Kunst des Zusammenmachens," in: Baier Andrea, et al., eds, *Die Welt reparieren: Open Source und Selbermachen als postkapitalistische Praxis*. Bielefeld: transcript, 2016: 34–62.

[8] Cf. the discussion in Möser, Kurt, "Thesen zum Pflegen und Reparieren in den Automobilkulturen am Beispiel der DDR," *Technikgeschichte* 79 (3), 2012, 207–226.

[9] Vasilyeva, Zinaida, *From Skills to Selves: Recycling "Soviet DIY" in post-Soviet Russia*. Neuchâtel: University of Neuchâtel, 2019.

[10] Golubev, Alexey, and Olga Smolyak, "Making Things through Making Things: Soviet Do-It-Yourself Culture and Practices of Late Soviet Subjectivation," *Cahiers du monde russe* 54 (3–4), 2013: 517–542. This magazine *Modelist-Konstruktor* is still being published today.

implied by the term DIY. For this reason, "do it ourselves" is, to my mind, the most appropriate concept to use when analyzing these activities.[11]

In this chapter, I use the phrase "We did it ourselves" as the starting point for nudging History of Technology, as a field, toward a stronger focus on repair and maintenance activities, and toward the reality of collective work. Around the turn of the twenty-first century, historians and sociologists began to acknowledge the active incorporation of artifacts and systems into consumers' and users' lives. The slogan of the era was "Users matter."[12] Historians of technology began to realize that scientists and engineers, economists and politicians are not the only ones who bring about technical change—and that History of Technology cannot narrate inventions and innovation processes to the exclusion of users. By blurring the boundaries between users and producers, Ruth Oldenziel and I made a case for studying organized consumers and amateur "tinkerers."[13] In this chapter, I show how ordinary people in Uzbekistan and neighboring countries have taken on selected construction and maintenance projects *collectively*, hewing to an age-old Muslim way of organizing communal life. Although the examples in this chapter refer to Central Asia only, it is important to know that this practice of "We did it ourselves" extends to many parts of the world.

Most of the new empirical material presented in this chapter derives from interviews conducted for the research project on which this book is based. The interview excerpts included here highlight the relevance and value of oral history to contemporary history as a field. In pursuing their research, my colleagues who contributed to this chapter adopted established oral-history methods. Given their need to understand events in interviewees' distant pasts, the researchers gave subjects ample time for reflection. And given that interviewers were often asking personal questions, it was vital for them to build mutual trust. This meant that the

[11] The analysis in this chapter, including the elaboration of the "We did it ourselves" concept, derives mainly from the work done by (and discussions with) Jonas van der Straeten and Mariya Petrova under the auspices of the ERC project mentioned in Footnote 1. Cf., e.g., Straeten, Jonas van der, and Mariya Petrova, "Mud Bricks in a Concrete State: Building, Maintaining and Improving One Own's House in Soviet Samarkand, 1957–1991," in: Krebs, Stefan and Heike Weber, eds, *The Persistence of Technology: Histories of Repair, Reuse and Disposal*. Bielefeld: transcript, 2021: 111–137.

[12] Oudshoorn, Nelly, and Trevor J. Pinch, eds, *How Users Matter: The Co-construction of Users and Technology*. Cambridge, MA: MIT Press, 2003.

[13] Oldenziel and Hård, *Consumers*.

interviews were relatively long, and that they did not follow a preset-questionnaire model.[14]

To provide readers with more examples of collective repair, maintenance, and construction work, I include references to, and quotations from, further interviews conducted by fellow historians and anthropologists. Considered as a body of evidence, this broader collection of interviews can be read as an important contribution to contemporary history writing, especially to the established practice of documenting the history of everyday life: "writing history from below."[15]

State Power and Beyond

Members of my research team asked informants to describe their housing arrangements in Samarkand during and after Soviet rule. This direct inquiry into residents' activities, "from below," so to speak, was meant to counterbalance the standard depiction of people in Uzbekistan as passive recipients of governmental housing and other public services. By interviewing dwellers of individual homes, for example, the historians sought to avoid an undue emphasis on people living in large-scale, state-run, functionalist areas, so-called microdistricts (*mikroraiony*). This choice of interviewees enabled the researchers to discover the extensive individual and cooperative action that took place—even during the Soviet era.

During the Cold War, most Western historians subscribed univocally to the one-sided view of the Soviet Union as a cruel power-machine. Specialists in the history of Central Asia under Soviet rule tended to reproduce this narrative, and they traced the roots of subjugation to Czarist times.[16] In the 1990s, when archives became more readily available, somewhat more diverse narratives developed. One habit remains, however: scholars of history and political science tend to extrapolate the circumstances of the infamously brutal Stalinist phase to apply to other periods. For example, historian Sergey Abashin criticizes political scientist James C. Scott for drawing general conclusions about all of the Soviet Union—and the entire Soviet period—from source material relating only to the

[14] Concerning the theories and methods of oral history, see Perks, Robert, and Alistair Thomson, *The Oral History Reader*, 3rd ed. London: Routledge, 2016.

[15] Cf. Chap. 1 and Thompson, Edward P., *The Making of the English Working Class*. London: Gollancz, 1963.

[16] Allworth, Edward, ed., *Central Asia: A Century of Russian Rule*. New York: Columbia University Press, 1967.

European parts of the Russian Soviet Federative Socialist Republic (RSFSR) during the Stalinist era.[17]

Scott's Soviet case study, published in his book *Seeing Like a State*, examines the collectivization and socialization of the land and the creation of large-scale agricultural units. Scott analyzes these processes by means of the concept "high modernism." According to Scott, high-modernist reformers believed that replacing individual farmland with collectively run *kolkhozy* and state-run *sovkhozy* would rationalize agricultural production and lead to higher crop yields. Scott argues that the practical outcome of the reformers' ideology—and the state's ambitions—was disastrous: peasants who managed to survive the Great Purge found themselves to be nothing more than modern-era serfs. Without land of their own, peasants were forced to accommodate to the rules of collective farms—or to become state employees.[18]

Abashin's portrayal of collective farm life in the Uzbek Soviet Socialist Republic contrasts sharply with Scott's analysis. Whereas Scott sees his subjects as victims of state repression, Abashin, in his research on a small village in the post-Stalin era, depicts enthusiastic people with "hopes and plans for the future." His "microanalysis of the local situation" reveals that *kolkhoz* members exercised considerable freedom to grow their own fruits and vegetables, and to sell their produce at the local market. From an Uzbek viewpoint, the *kolkhoz* model seemed to harmonize with traditional ways of life.[19]

Like Scott, who is a political scientist, historians of technology are also prone to associating Soviet power with large-scale infrastructure projects. In a broad-ranging article, US historian Paul R. Josephson analyzes the ideological underpinnings of "Lenin's electrification, Stalin's canals and hydropower stations, Khrushchev's atomic energy, and Brezhnev's Siberian river diversion project."[20] For Josephson, it was the monolithic

[17] Abashin, Sergey, "'Ideal'niy kolkhoz' v sovetskoi Sredney Azii: Istoriya neudachi ili uspekha?" *Acta Slavica Iaponica*, 29, 2011: 1–26, here: 2; cf. Straeten, Jonas van der, "Borderlands of Industrial Modernity: Explorations into the History of Technology in Central Asia, 1850–2000," *Technology and Culture* 60 (3), 2019: 659–687; here: 666.

[18] See Scott, James C., *Seeing Like a State: How Certain Schemes to Improve the Human Condition Have Failed*. New Haven, CT: Yale University Press, 1998, Ch. 6.

[19] Abashin, "'Ideal'niy kolkhoz,'" pp. 3–4.

[20] Josephson, Paul R., "'Projects of the Century' in Soviet History: Large-scale Technologies from Lenin to Gorbachev," *Technology and Culture* 36 (3), 1995: 519–559; here: 519.

power structure of the USSR that made it possible to mobilize enough resources to turn modernist ideas of technological grandeur into practical projects.

Historians interested in Soviet urban planning and city building also tend to emphasize the gigantism of many projects. Josephson's own monograph on Akademgorodok—which frames this Siberian city as the realization of a modern, Baconian utopia—is a case in point.[21] A substantial amount of literature exists on the history of so-called socialist cities, *sotsgoroda*.[22] Much of this literature focuses on the foundation and development of large, industrial, mono-functional cities in the socialist bloc. Examples include the Soviet-Russian cities of Uli-Ilimsk, built around a hydroelectric plant, and Angarsk, the site of large-scale petrochemical factories.[23] In Eastern Europe, the "steel towns" of Eisenhüttenstadt and Nowa Huta have garnered excessive attention from urban and economic historians.[24]

Indeed, in characterizing socialist cities, most historians emphasize their uniformity and drabness, as well as their tendency to embody rationalist planning ideals. Detailed case studies of daily life in such cities often provide a more balanced picture, however. For example, in his study of Magnitogorsk, a city designed to extract and process the abundant iron ore in the Ural Mountains, historian Stephen Kotkin provides an insightful counter-narrative to the standard-issue history of industrial cities in the USSR. Although Kotkin acknowledges that Magnitogorsk was planned as an urban utopia, his focus is not on planners' and architects' attempts to create an "ideal city." Instead, Kotkin brings out the "creative resistance" of ordinary citizens and unearths the activities of "individuals with hopes,

[21] Josephson, Paul R., *New Atlantis Revisited: Akademgorodok, the Siberian City of Science*. Princeton, NJ: Princeton University Press, 1997.

[22] This literature spans from French, R.A., and F.E. Ian Hamilton, eds, *The Socialist City: Spatial Structure and Urban Policy*. Chichester: John Wiley, 1979, to Meerovich, Mark G., Viachaslau I. Menkouski, and Igor L. Zherebtsov, *"Socialist City": Idea and Its Realization in the Soviet Union 1920's and 1930's*. Belianum: Banská Bystica, Slovakia, 2019.

[23] Engel, Barbara, *Öffentliche Räume in den blauen Städten Russlands: Entwicklungen, Status und Perspektiven; funktionale und räumliche Anforderungen an die Erneuerung öffentlicher Räume in den neuen Industriestädten Sibiriens unter veränderten sozioökonomischen Bedingungen*. Tübingen: Wasmuth, 2004.

[24] Jajeśniak-Quast, Dagmara, "In the Shadow of the Factory: Steel Towns in Postwar Eastern Europe," in: Hård, Mikael, and Thomas J. Misa, eds, *Urban Machinery: Inside Modern European Cities*. Cambridge, MA: MIT Press, 2008: 187–210.

fears, a capacity for survival, and no small amount of inventiveness."[25] By focusing on Samarkand and other Central Asian cities, I strive to do the same in this chapter.

The Persistence of the Neighborhood Community

We also find a tension between monotonous utopias and local realities in the literature on Central Asia. In his history of the Uzbek city of Tashkent under Stalin and Khrushchev's rule, scholar Paul Stronski analyzes the dialectics of modernist plans and indigenous traditions.[26] After Tashkent in 1930 was made the capital of the Uzbek Soviet Socialist Republic, architects and engineers, working on behalf of the State, began to turn the city into a showcase of modern civilization and socialist ideals. Representative buildings, wide avenues, large squares, parks, and factories were to replace mosques, narrow alleys, and what the modernists considered to be filthy, unhygienic residential areas. Considerable resources were mobilized to turn Tashkent into a model town for Soviet Central Asia.

The Communist Party was the driving force behind these urban developments. In their efforts to create both a "socialist city" and the "Soviet Man"—a person who would incorporate communist ideals of collective action and unselfish behavior—the Party tried to erase traditional structures, both material and social.[27] In 1954, the Central Committee of the Uzbek Communist Party stated this goal forcefully: "We must liquidate the mud-brick home, we must liquidate the Old Uzbek way of life."[28] By removing traditionally designed courtyard houses and forcing people to move to European-style residential areas, Party officials aimed to make them accustomed to "modern" ways of life.

The modernizers never managed to realize all their goals. Although Tashkent became one of the most Sovietized cities in the region, Stronski concludes that "the traditional Central Asian neighborhood [...], with its serpentine streets, internal courtyards, and local customs, continued to prove much more resilient than many urban planners had initially

[25] Kotkin, Stephen, *Magnetic Mountain: Stalinism as a Civilization*. Berkeley, CA: University of California Press, 1995, pp. 19, 154.

[26] Stronski, Paul, *Tashkent: Forging a Soviet City, 1930–1966*. Pittsburgh, PA: University of Pittsburgh Press, 2010.

[27] Ibid., p. 63.

[28] Quoted from ibid., p. 308.

envisioned."[29] Having learned how to subvert the Soviet legal system, individual house owners often successfully resisted eviction and confiscation. Even when ultimately forced to move, many residents built new houses with traditional layouts on the outskirts of town.[30]

One reason for the refusal to move into the modern apartments of multistory buildings was the inappropriate design of the housing. The standard new apartments were intended for nuclear families consisting of two grown-ups and two children, rather than traditional extended families. The government-sponsored apartments also had a design flaw: during the excessively hot summer months, the homes became practically uninhabitable. An additional problem was that most residents of large, reinforced-concrete buildings had no access to land, and they were unable to cultivate fruits or vegetables to meet their personal needs.

Facing resistance, architects tried to make life in modern residential areas more appealing. One idea was to create an area in Tashkent's Old Town that consisted of modern buildings of different sizes and would provide some families with access to courtyards and rooftop gardens. To underscore that the project was meant to allow the "native population" to continue living close to relatives and extended families, the architects called it "Mahalla"—a reference to traditional Islamic communities.[31] Before the project could be realized, however, Tashkent in 1966 was hit by a devastating earthquake, after which the planning changed considerably.

The word *mahalla* can be found in Uzbek and other Turkic languages, as well as in Arabic and Persian. Often translated as "neighborhood community," a *mahalla* does not simply denote a group of people living in close proximity to each other, however. A *mahalla* has both a spatial and a social dimension, and it is mostly organized around a mosque. It is common to find a public water source, and perhaps a tea house in a *mahalla*. Traditionally, Central Asian *mahallas* were managed by an elected official, who organized community work, functioned as a bridge to the city or

[29] Ibid., p. 269.

[30] Petrova, Mariya, *"Nah am Boden": Privater Hausbau zwischen Wohnungsnot und Landkonflikt im sowjetischen Samarkand der 50er und 60er Jahre*. Berlin: De Gruyter, 2020.

[31] Stronski, *Tashkent*, pp. 232–233. Stronski, on p. 232, has taken the quote about the "native population" from a 1968 article by Z.N. Chebotareva in the journal *Stroitel'stvo i arkhitektura Uzbekistana*.

national government, and collected taxes.[32] Some *mahallas* initially brought together members of a certain guild or a particular ethnic group but later diversified. In Tashkent, for example, the Mahalla Ukchi used to be the neighborhood of the weapon blacksmiths, and the Tajik-Mahalla originally attracted people of Tajik descent.[33] Funerals, weddings, and other communal events brought together members of a *mahalla*. Importantly, *mahallas* carried out essential support functions in societies lacking a state-run welfare system.[34]

Despite their communal character, *mahallas* also enable a high degree of social control on the local level. Some Soviet Republic and post-Soviet governments appropriated this age-old institution to make it suit their own purposes—for coercion and surveillance, for example. In today's Uzbekistan, the *mahalla* concept has been incorporated into the official local administration, and the government has even created new *mahallas* in modern residential areas.[35]

Construction Materials and Climate in Central Asia

The courtyard house is a traditional design in Muslim regions—beyond Central Asia as well. The courtyard itself typically includes a small vegetable garden, a few trees, and, sometimes, a well. Many activities take place in the yard—especially in the summer. Measures are taken to ensure that such activities are not visible to the outside world. Unlike the residential buildings in most European towns, residential structures in so-called Islamic cities often lack windows facing the street or the alleyway. High walls surrounding the courtyard manifest the separation between public and private life. In the private sphere, it is common to find additional

[32] Faizulloev, Muhiddin, "Mahalla in Northern Tajikistan from the End of 19th to the Beginning of the 20th Century: The Case of Khujand and Its Suburbs," in: Roche, Sophie, ed., *The Family in Central Asia: New Perspectives*. Berlin: Klaus Schwarz Verlag, 2017: 55–83.

[33] Gangler, Anette, et al., eds. *Städte Usbekistans zwischen Tradition und Fortschritt: Städtische Transformationsprozesse der zentralasiatischen Städte Taschkent und Samarkand*. Cottbus and Stuttgart: BTU Cottbus and Stuttgart University, 2006, p. 26.

[34] Jordan, Jens W., "Die Mahallas der timuridischen Altstadt Samarkands," in: *2750 Jahre Samarkand*. Erfurt: Stein, 2007: 58–65.

[35] Pétric, Boris Mathieu, "The Mahalla in Present-day Uzbekistan," in: Gammer, Moshe, ed., *The Caspian Region, Volume I: A Re-emerging Region*. London: Routledge, 2004: 228–243; Alexander, Catherine, and Victor Buchli, "Introduction," in: Alexander, Catherine, Victor Buchli, and Caroline Humphrey, eds, *Urban Life in Post-Soviet Asia*. London: University College London Press, 2007: 1–39, here: 29.

forms of separation: between traditionally defined female and male areas; between areas where guests are entertained, and rooms to which only family members have access; between buildings inhabited by the older generation, and those designated for one of the sons and his family; and between living areas used in winter, and those used only in summer.[36]

Courtyard houses in Uzbekistan were traditionally built with loam bricks. These bricks were often made of a particular aeolian clay called loess—a mixture of sand, silt, and clay. Loam bricks were used for walls that faced the street as well as for various buildings within the courtyard. The advantages of using loam are clear: loess is abundant throughout Central Asia, and the making of unfired bricks with this material is a low-tech activity. In this sense, adobe (unfired bricks) is an excellent DIY building material, requiring neither large amounts of capital nor extensive engineering knowledge. When the construction area is large enough, bricks may be made on site. In any case, to make bricks, the builder or mason mixes loess with water and straw, then pours this mixture into brick-shaped molds. Ultimately, the bricks are laid out to dry in the sun. The product is commonly called mud bricks; in English, the word *adobe* (from Spanish) has gained currency.

The use of adobe—whether in the form of plaster or unfired bricks—presents a trade-off. One advantage is that the material for making adobe is inexpensive and often easily accessible. Another benefit is the relative flexibility of adobe as a building material: given that it is not as stable as, say, concrete, adobe structures lend themselves to being "disassembled" when needed. For example, when a family needs to remove walls in order to build an extension, mud-brick walls are far easier to demolish. Accordingly, the drawback of an adobe-brick wall is that it may not last as long as one made of reinforced concrete; also, adobe requires ongoing maintenance.

To document the traditional brick-making process, researcher Enrico Fodde interviewed artisans in various Central Asian countries. Fodde's interviews, conducted in the early 2000s, document processes that have likely remained unchanged for much of the last century. (Evidence points to the arrival of Czarist troops—and Russian influence—affecting

[36] Bianca, Stefano, *Hofhaus und Paradiesgarten: Architektur und Lebensform in der islamischen Welt*. Munich: C.H. Beck, 1991, pp. 208–244.

brickmaking techniques in the nineteenth century.[37]) When large quantities of bricks were needed, masons preferred sites close to rivers or canals. Here, loam is often plentiful, and access to water is practically guaranteed. Once a site had been located, the craftsperson tested the texture of the mud. If the quality was acceptable, workers dug out an appropriate area with a mattock (a combination of pickaxe and adze), crushed larger pieces of clay, and added water. When they lacked access to machinery, workers mixed loam and water using their bare feet. Small amounts of straw or other binding material were added. The brick molds used were often made of wood, and they were provided by carpenters. In earlier eras, bricks were square; later, rectangular bricks (also known as "Russian bricks") became more common.[38]

The climate in Central Asia has contributed to the widespread use of loam as a construction material. The region is characterized by extreme temperature differences: very hot summers and bitter-cold winters. Substantial variations may also occur between day- and nighttime temperatures. When it comes to creating stable indoor temperatures, thick loam walls—and the use of loam in roofs—do a better job than wood. Fodde describes Tajik village roofs made of a wooden frame covered by a thick layer of reed, bark, straw, and as much as 30 centimeters of clay.[39]

A drawback is that loam walls are affected by weather erosion and wear-and-tear from residents. To prevent deterioration, walls may be plastered—or at least whitewashed. Typically, clay-based plaster includes a larger amount of straw to prevent the surface from cracking. Animal feces may also be added.[40] Plastered walls are typically renovated every ten or twenty years, whereas whitewashed walls need to be maintained at five-year intervals.[41]

[37] Pisarčik, A.K., *Narodnaya arkhitektura Samarkanda: XIX-XX vv.* Dushanbe: Izdatel'stvo "Donish," 1974, p. 57.

[38] This paragraph is based on Fodde, Enrico, "Traditional Earthen Building Techniques in Central Asia," *International Journal of Architectural Heritage: Conservation, Analysis, and Restoration* 3 (2), 2009: 145–168. Cf. also Pisarčik, *Narodnaya arkhitektura*, Ch. II. Concerning the concept "Russian bricks," see Cooke, Louise, "Earthen Building Materials and Techniques at Merv, Turkmenistan," in: *LEHM 2004: 4th International Conference on Building with Earth.* Weimar: Dachverband Lehm, 2004: 52–61; here: 57.

[39] Fodde, "Traditional," pp. 160–162.

[40] Ibid., pp. 158–160.

[41] Cooke, "Earthen," p. 57.

Given the extreme weather in Central Asia—heavy precipitation, harsh winds, and scorching sunshine—roofs that incorporate loam must be maintained regularly. For example, when inhabitants remove snow from flat roofs in winter, the top layer of loam may sustain damage. Rooftops are also used as a storage area and as a playground for children; both activities add to the wear-and-tear on the structure. Fodde's Tajik interviewees from the Zerafshan area informed him that, every five years or so, they add a couple of centimeters of clay to the top of their roofs.[42]

Climatic conditions also limit the window of time for building construction in Central Asia. Adobe bricks can only be made during the hot, dry summer months. House building starts in the spring and may take place until November, but seldom longer. By and large, construction used to be a male activity. Women sometimes played a role in plastering and whitewashing walls; evidence from Turkmenistan shows that women were often responsible for painting and wallpapering.[43] The homeowner's decision about whether to engage experienced builders was based on the family's finances. As a rule, however, family members and relatives took an active role in the construction process, whether in building from scratch or helping with renovations. Neighbors and other members of the village or the *mahalla* also lent home builders a hand.

In a conversation with anthropologist Victor Buchli in the early 2000s, an interviewee named Auntie L. explains the advantages of adobe bricks. Auntie L. lived (and still lives, one hopes) with her husband in the originally Czarist part of Astana (called Tselinograd in Soviet times and Nur-Sultan between 2019 and 2022), the modern capital of Kazakhstan. At the time of the interview, Auntie L. had recently replaced a dilapidated part of her house with a new section, intended for the youngest son and his family. For this extension, she had decided to use mud bricks, which promised to protect against the cold temperatures and gale-force winds of the Kazakh steppe. Auntie L.'s land had plenty of loess soil, making it easy to produce adobe bricks on-site. The entire extended family took part in the construction process. They mixed loam with water and hay, and subsequently, they set out the bricks to dry in the sun. The men in the family raised the walls in as little as three days, and the women plastered the walls

[42] Fodde, "Traditional," p. 162.
[43] Cooke, "Earthen," p. 55.

with a layer of clay. Together, the family members built the addition to the house-themselves.[44]

One of the people my research colleagues interviewed testified to the simplicity of the adobe-manufacturing process. As a teenager, Bakhrom worked with friends during the summer months to make bricks at various locations. Neither specialist knowledge nor advanced equipment was required; Bakhrom and his friends mixed the loam with water, formed the bricks, and lay them in the sun to dry. To save time and money on travel, the friends built themselves a small hut close to each production site, where they spent the evenings cooking before they slept.[45]

Another interviewee, Wladimir, remembers his parents hiring a pair of workers to make adobe bricks, from loam, in their own yard: "First they came to have a look and walked around the yard." After the workers found suitable soil, "they began to shovel. After removing the top layer, they dug deeper. When one of them found a good layer, they extracted the loam [...] and added some sand." The loam-sand mixture was then thrown back into the hole, and water was added. After a couple of days, when the mixture had taken on the right texture, the workers started forming the bricks. Working "quickly," they managed to make several hundred per day. At the end of this process, the adobe bricks were placed in the hot sun to dry for approximately one week. "Most of the additions and other things" in the family's yard were eventually built with these bricks.[46]

Between *Hashar* and *Subbotnik*

We know that people throughout the centuries have come together to undertake tasks of common concern: constructing houses of worship; repairing roads; combatting fires; and designing irrigation systems. Communal work was, and may still be, especially urgent in regions with a low population density (as in the North American countryside), a lack of specialists (as in distant Norwegian valleys), and a weak monetary

[44] This paragraph is based on Buchli, Victor, "Astana: Materiality and the City," in: Alexander, Catherine, Victor Buchli, and Caroline Humphrey, eds, *Urban Life in Post-Soviet Asia*. London: University College London Press, 2007: 40–69, here: 56.

[45] The information in this paragraph is taken from an interview with Bakhrom (name changed), carried out by Mariya Petrova and Jonas van der Straeten on September 18, 2018.

[46] The information and quotations in this paragraph are taken from two interviews with Wladimir (name changed) carried out by Mariya Petrova and Jonas van der Straeten on June 19 and July 29, 2018.

economy (as in many Central Asian areas).[47] People also gathered to help individuals carry out tasks they could not manage on their own. For example, in a ritual called barn raising, rural farmers in the United States would collaborate to help a neighbor build a barn; similarly, Uzbek villagers would congregate to help a neighbor repair a leaking roof.[48]

Communal work has a long history in Central Asia. Bearing certain similarities not only to barn raising, but also to *dugnad* (voluntary work for a common good) in Norway, this social practice is widespread both in the countryside and urban neighborhoods.[49] In Tajik, this form of mutual aid is called *hashér*; Kazakhs employ the term *assàr*, and Uzbeks use the word *hashar*.[50] The fact that their root lies in the Arabic word *hšr*, meaning community or collective, indicates this is an old institution in Turkic, Persian, and Arabic communities alike.[51] In an interview with anthropologist Christilla Marteau d'Autry, one inhabitant of Samarkand defined *hashér* as an event in which "everyone gathers for the benefit of society," to "work together."[52]

During Soviet times, the State introduced the idea of the *subbotnik* across the entire USSR. In Russian, *subbota* means Saturday; the idea was that Soviet citizens should devote at least one Saturday per year to unpaid, communal work. To a certain extent, this attempt by the State to mobilize local communities for its own purposes reminds us—as we saw in Chap. 7—of the ways in which the Kenyan authorities used the concept of *harambee* ("Let's pull together") in various campaigns.

Immediately after the fall of the Soviet Union, the independent Uzbek government reinvented *hashar* to replace the Communist *subbotnik*. By transforming the *hashar* tradition into an Uzbek alternative to the mandatory Soviet Saturday, the authorities tried to kill two birds with one stone: to boost solidarity, and to motivate citizens to undertake important

[47] For a general overview of this kind of activities, see Lorentzen, Håkon, and Line Dugstad, *Den norske dugnaden: Historie, kultur og fellesskap*. Kristiansand: Norwegian Academic Press, 2011, pp. 9–41.

[48] Marteau D'Autry, Christilla, "*Vyjdem vse, kak odin!* 'Allons-y tous comme un seul homme!' Ethnographie d'un *hashar* national dans un quartier de Samarkand, Ouzbékistan," *Cahiers d'Asie centrale* 19–20, 2011: 279–301; here: p. 281.

[49] Concerning this long-standing practice in Norway, see Lorentzen and Dugstad, *Dugnaden*.

[50] Fodde, "Traditional," p. 153.

[51] Marteau D'Autry, "*Vyjdem vse*,", p. 279.

[52] Ibid., pp. 287, 292.

communal tasks. Such work is often carried out as a part of the preparations for *Nowruz*, an official holiday celebrated on March 21, the day of the vernal equinox. On the local level, the official *mahallas* are responsible for organizing these *hashar* events.[53]

In 2007, Marteau d'Autry attended a "national beneficence *hashar*" in a Tajik *mahalla* in Samarkand. Situated on the outskirts of the city, the *hashar* activities in this *mahalla* included various tasks. In addition to cleaning streets and painting the community center, participants were expected to clean the water basin and other areas of the old cemetery; to plant trees; to lay the foundation for a mosque; and to start construction on a sports field. Women were explicitly asked to sweep streets and alleyways, and children helped to clean the grounds of the local cemetery. Other tasks were carried out by both unskilled and skilled men. Like everyone else, a professional carpenter, electrician, and house painter contributed their labor without remuneration—although the *mahalla* reimbursed them for any materials. The elderly attended as well, drinking tea while offering their knowledge and experience to the younger generation. Marteau d'Autry's somewhat romantic story paints a picture of *hashar* as a social event that contributes to keeping the neighborhood together—while reproducing gender roles, generational differences, and status symbols. The collective work typically ends with common meals or festivities.[54]

The extension and maintenance of private buildings may also take place *hashar*-style. In this case, rather than calling on the entire *mahalla*, owners mobilize family members, close or distant relatives, and, perhaps, several direct neighbors.[55] Mukhabbat, an interviewee living on the outskirts of Samarkand, confirmed this practice. In 1986, when Mukhabbat and her husband wanted to build a dining hall large enough to accommodate at least thirty-five people, the couple relied on their relatives. "We have always had the *hashar* method. For example, we have a big family—brothers come, my nephews [...]." In exchange for their work, the team members received three meals per day: "We had a good breakfast [...], for lunch there had to be pilaf or cabbage rolls." After a whole day's heavy

[53] This and the following paragraph are based on ibid.

[54] Scholars criticize overly romantic images of *hashar*, pointing out that—after the fall of the Soviet Union—well-off persons would hire others to do the required tasks for them; cf. Koroteyeva, Victoria, and Ekaterina Makarova, "Money and Social Connections in the Soviet and Post-Soviet Uzbek City," *Central Asian Survey* 17 (4), 1998: 579–596.

[55] Faizulloev, "Mahalla," p. 72.

work, everybody enjoyed *chorba*, a hearty soup.[56] Mukhabbat's teenaged nephews received money for their work, in addition. The nephews were assigned the task of making the loam bricks needed for the sixty-square-meter dining hall. Like the workers in Wladimir's narrative, Mukhabbat's nephews made the bricks on-site.

Repair and Maintenance

All dwellings are affected by the ravages of time; people across the world are compelled to repair leaky roofs and water pipes, deteriorating walls and window frames. What varies, though, is the amount of work we commit to home maintenance. For example, the most active homeowners arrange for their house to be painted—or whitewashed—on a regular basis; other residents, whether for lack of time, funds, or motivation, may simply keep their fingers crossed, in the hope that the effects of time will not destroy their abode. An engineer once calculated that almost one-third of the money spent on construction in Britain covered repair and maintenance work.[57]

Historians of technology have only recently discovered the importance of repair and maintenance.[58] After decades of researching inventions and innovations, members of the profession have finally become curious about what happens to technological structures as they age. In the last decade, anthropologists have also begun to investigate how communities manage the built environment. Interviews and historical sources reveal how people in Central Asia combat decay—and take precautions for the future.

Scholar Wladimir Sgibnev documents the broad scope of repair and maintenance activities in parts of Central Asia. During his fieldwork in 2009 and 2010, Sgibnev interviewed homeowners and tenants in the Tajik city of Khujand (called Leninabad during the Soviet era). At the center of his analysis is *remont*, an originally French word that entered Central Asian languages from Russian. Like the English words renovate (a transitive verb) and renovation (the noun), *remont* captures various kinds of repair and maintenance activities, as well as reconstruction work.

[56] The quotations in this paragraph are taken from an interview with Mukhabbat (name changed), carried out by Mariya Petrova and Jonas van der Straeten on September 22, 2018.

[57] Seeley, Ivor H., *Building Maintenance*. London: Macmillan, 1976, p. 2.

[58] Russell, Andrew L., and Lee Vinsel, "After Innovation, Turn to Maintenance," *Technology and Culture* 59 (1), 2018, 1–25.

Remont can be used as a verb and as a noun.[59] One homeowner Sgibnev interviewed used the word *remont* in its most versatile, encompassing form:

> We had an old house but we made remont there. It's a normal Tajik house, but now there is a remont like in Europe. [...] We have done all the remont already, it looks like new. But [...] we will do another remont when we have a wedding. Then we will renew the remont.[60]

The interviewee indicates that substantial work went into repairing the original building, which had been in poor condition. One of the goals of the renovation was to adopt European standards, which were apparently considered superior to Tajik ones. Well aware that any building must be maintained regularly, the owner claimed that his family would undertake another round of renovations in preparation for a future marriage.

In addition to showing that recurring maintenance was crucial for loam buildings, Sgibnev documents how people in modern concrete-panel buildings carried out repair and renovation. Often, this was easier said than done, however. In this interview, an apartment owner describes some of the difficulties with living in a building designed during the Soviet period:

> You've got this problem in every house. In this flat which I bought, too, the roof is leaking. I was told already that I had to fix it myself. So I will start with the roof, and then go on to a *remont* of the apartment.[61]

Other anthropological studies also discuss maintenance activities in multi-story, concrete buildings. During his field work in 2008–2009, Mateusz Laszczkowski observed the problems faced by residents of an apartment block in Astana. Residents could trust neither service providers nor the authorities to carry out the necessary repairs, so they had no choice but to address maintenance work themselves, repairing the building's plumbing system and outdoor facilities, for example. Given the high turnover of residents in the 1990s and the 2000s, social cohesion in the building was weak; repair-and-maintenance initiatives depended on a small group of

[59] Sgibnev, Wladimir, "*Remont*: The Social Production of Space in Central Asia," *PhD Dissertation*. Berlin: Humboldt University, 2012.

[60] Quoted from Sgibnev, Wladimir, "Remont: Housing Adaptation as Meaningful Practice of Space Production in Post-Soviet Tajikistan," *Europa Regional* 22 (1–2), 2014: 53–64, here: 55.

[61] Quoted from ibid., p. 57.

active individuals. One of the most devoted residents told Laszczkowski, "No one wants to do the job."[62] Apparently, a certain tension between apartment owners, long-term tenants, and short-term tenants made it difficult to organize collective work.

Doing It Together: Cooperation and Collectivity

The interviews we conducted support the conclusions drawn by historians and anthropologists who have studied urban life in Central Asia. Our research also complements the existing body of knowledge: our interviewees describe how ordinary people actively designed, repaired, and maintained their living spaces—cooperatively and collectively, with family members and with neighbors.

Anvar, a Samarkand merchant, recalls how residents of an urban *mahalla* would gather on a yearly basis to "help each other." In the 1960s, when Anvar was still a child, the roofs of his parents' and neighbors' houses "did not have fiber cement covering, only clay." This solution was cheap and simple, but it had its obvious drawbacks. Anvar recalls:

> Sometimes, when we were sleeping, in the wintertime, water dripped from the roof (the roof was leaky), and we had to dash off to get coal ash, to sprinkle onto the roof. It turned into a kind of cement which stuck to the roof. We took a bucket and then sprinkled salt on top of it [...].

To avoid troublesome leaks, the roofs had to be repaired and maintained "every year." This work was usually done in September, before the first snow fell. Members of the *mahalla* organized the labor themselves: "We started with the tea houses, then your roof, his roof." It was "like *hashar*." No professional builders took part, and to motivate the volunteer participants, "a good meal at lunchtime with pilaf or vodka" was provided. Although these activities took place during the Communist era, they had nothing to do with the officially prescribed practice of *subbotnik*. Anvar describes collective work in which neighbors collaborated to solve problems of common interest—or to help each other.[63]

[62] This paragraph is based on Laszczkowski, Mateusz, "Scraps, Neighbors, and Committees: Material Things, Place-Making, and the State in an Astana Apartment Block," *City & Society* 27 (2), 2015: 136–159; here: 150.

[63] The quotations and information in this and the following two paragraphs are taken from an interview conducted by Mariya Petrova and Jonas van der Straeten on September 25, 2018, with Anvar (name changed).

Only in the early 1980s did residents of the *mahalla* get ahold of fiber-cement sheets with which to cover their roofs: “One neighbor began to bring asbestos sheets, and another asked where he had gotten them—and it went on and on.” Anvar worked at a gas station at the time; he recalls how he “sent a truck to the *mahalla* where they, day after day, removed 50 centimeters of clay from the roof and replaced it with fiber-cement sheets.” Asbestos sheets had the advantage of being both waterproof and considerably lighter than the thick layer of clay. If the fiber cement was of good quality, the sheets would also require less-frequent maintenance. (At the time, asbestos was not known to be carcinogenic.)

In Soviet times, the process of gaining access to construction materials differed from that of capitalist societies. The members of the *mahalla* managed to acquire asbestos sheets and beams at a local lumberyard for a very low price. To secure permission for this, however, the local Mahalla Committee was required to first submit a request to the Local Executive Committee of the Communist Party. Only after the Party Committee “had issued an order could you go there, pick up your sheets, and pay.” Individuals had to go through the same procedure every time they needed “bricks or boards or asbestos sheets.”

Bakhrom, who grew up in a large house on the outskirts of Samarkand, was lucky enough to have a close friend whose father was a high-ranking member of the Regional Committee of the Communist Party. Whenever Bakhrom wanted boards or fiber-cement sheets for his own house—which he built with his father, on his parents’ property—his friend made sure Bakhrom got what he needed. Via other personal connections, Bakhrom’s family was able to procure iron hardware. Nevertheless, it took Bakhrom and his father as long as eight years to finish the new house: “I received the beams, and then I had to wait for the boards. While making the floors, I had to wait for the laths.”[64]

Anvar and Bakhrom’s experiences were common ones. Another of our interviewees, Wladimir, relates that written permission from the Executive Committee was seldom enough to secure the necessary building materials: workers and security personnel at the lumberyard expected a gratuity, and the lumberyard’s truck driver had to make sure he had reserved enough gasoline to make the unofficial delivery—without getting into trouble with his superiors. In the 1980s, when Wladimir needed building material

[64] The quotations and information in this paragraph are taken from the interview mentioned in Footnote 45.

to create a footpath in his yard, he managed to get his hands on three bags of powdered cement from the company for which he worked—by tipping one of the foremen.[65]

Like many others in former socialist republics, Anvar remembers the past in a positive—or even nostalgic—way:

> At that time, people respected each other. Cultural standards were very high. And today, nobody needs anyone. Back then, if our neighbor was doing something, my father would come and ask me if we shouldn't make something, too.

In Anvar's opinion, the traditional urban *mahalla* gave people a particular sense of belonging that could not be found in the modern concrete-panel buildings of the city: "I grew up in the *mahalla*; it taught us children everything." Through experience, customs, and daily "practice," children in the *mahalla* learned what was expected of them as grown-ups. Weddings and funerals drew people together.[66]

Anvar grew up in low-income circumstances, which he expresses as, "We were ordinary people." His family lived in a typical courtyard house in Samarkand. Today, the house's "old doors" and nicely "carved columns" attract curious tourists who may well have a romantic view of life in the old town. The reality was far from romantic: not only did the roofs leak; one of the rooms in Anvar's house was "very cold" in the winter, and the kitchen walls were made of thin plywood boards, just enough to "protect it from the wind." As Dilya also told us, no one in the Soviet era really valued the old architectural style.[67] For example, Anvar's parents had the old columns in their home "covered with plywood."

Heating technologies reflected the humble living conditions. In the early 1970s, before gas pipes were installed in the *mahalla*, residents used masonry stoves and *sandals*—small pans or pits, which were filled with pieces of smoldering wood or coal. A *sandal* is a truly simple heat source, usually found underneath dining tables.[68] The slowly burning substance is

[65] The information in this paragraph is taken from the interviews mentioned in Footnote 46.

[66] The quotations and information in this paragraph are taken from the interview mentioned in Footnote 63.

[67] Cf. the interview mentioned in Footnote 1.

[68] Cf. El Hariry, Shorouk, et al., "Toward a Global History of Material Culture," *Technikgeschichte* 88 (2), 2021: 178–182.

simply put in a small hole in the ground or in a small pan on the floor. Anvar remembers the small pit under his grandmother's dinner table, where red-hot pieces of coal were placed to keep everyone's feet warm. Central Asian dining tables are low, and people used large tablecloths or carpets to hang well below the table's surface; this made the *sandal* a highly efficient heating technology.

Dilya, too, mentions that her grandfather's house was equipped with masonry stoves, and she also recalls the importance of the small *sandals*:

> Coal was placed under the table. The table was completely covered with a rug, and you would then put your legs underneath. In Afghanistan, they also had this type of fire pit. It was a very smart thing. In the winter, it was even possible to keep a small room warm with it.

"Out of habit," the family held on to some of their coal pits after their home was connected to the city's gas network.[69]

Masonry stoves are called "Dutch stoves" (*gollandskaja pechka*) in Russian, and most likely they were introduced into Central Asia during Czarist times. Anvar relates that during his childhood, family members had managed to build their own masonry stoves. Only when the Soviet administration introduced gas was it necessary to hire specialists to insert gas burners into the existing masonry stoves: "The experts got twenty-five rubles for their work, and they worked for two days. And then the gas was connected." As long as the residents were not hooked up to any technological infrastructure, they "did not need any experts" or craftspeople. In the extended family, there were always enough "brothers" who "knew something."[70]

Tatiana, first introduced at the start of this chapter, tells a similar story. Tatiana grew up in Samarkand, in the kind of standardized private house that residents built themselves—a kind of State-supported do-it-ourselves activity. This housing was part of an official construction program of the late 1950s and early 1960s. Born in 1950, Tatiana recalls that, in the beginning, her family only had a bathroom and a toilet in the yard. To

[69] The quotations in this paragraph are taken from the interview mentioned in Footnote 1. Mukhabbat, in the interview mentioned in Footnote 56, describes the *sandal* in a similar way.

[70] The quotations in this paragraph are taken from the interview mentioned in Footnote 63.

fetch water, family members also had to go outside. The house had a masonry oven, but in the mid-1960s, her parents decided to switch to gas for heating and cooking. Electricity had been available early on.[71]

Making the Comfortable Home

Gas pipes were only one of the technical systems that found their way into Soviet Uzbekistan. The history of electricity provision goes back to the interwar period, and in the same era, some urban areas began to receive piped water. The house where Dilya grew up was connected to the urban gas network in the late 1960s. Initially, family members were skeptical about gas heating: "We don't need it, we don't know what that thing is." Dilya explained their reluctance: "You know, we had the small, coal-fired *sandal* ovens. They were cheap, and people were used to them." But when some of Dilya's neighbors decided to get a gas connection, her grandfather decided to follow suit. Apparently, the fact that gas was considered new and modern justified its high cost.[72]

The development of large-scale, modernist construction projects went hand in hand with the introduction of large-scale, centralized technical systems. Access to essential services facilitated life in multistory residential housing complexes. From their inception in the late 1950s, concrete-panel building areas were designed to provide electricity and water, as well as sewerage, district heating, and gas.

Natalya, another one of our interviewees, has fond memories of when she and her parents moved into one of Samarkand's newly built apartment blocks, in 1976:

> Absolutely everything was there, electricity and light bulbs, wood everywhere, everything was painted, there was a toilet and a bathroom, a gas stove. Absolutely everything was there. Radiators, central heating, the radiators were hot. Everything was OK. [...] It was very comfortable in the winter; [the temperature inside] was up to 30 degrees, everything worked, the co-generation plant worked without problems. We turned on the heat at the end of October; there was always hot water; there were never any problems.

[71] The quotations and information in this paragraph are taken from the interview mentioned in Footnote 3.

[72] The quotations and information in this paragraph are taken from the interview mentioned in Footnote 1.

When the family moved into their new apartment, Natalya was eleven years old. Her father was a somewhat high-ranking soldier, and the family was given a relatively large apartment: sixty-one square meters with three rooms and two balconies. The family bought beds and closets in a furniture store, and they asked a local carpenter to build a few items. Although the store's selection was limited, it reflected the transnational networks of the socialist block. Some products had been imported from the German Democratic Republic; others were "made in Poland."[73]

Echoing Anvar's nostalgia for the Soviet era, Natalya, a seamstress by vocation, expressed her satisfaction with the service providers of the 1970s and 1980s. If the sink was clogged, for example, Natalya remembers that she would simply call the Housing and Maintenance Office, and they would come by to fix it. Serious problems seem to have surfaced only in the late 1980s, when utopian visions of a comfortable life began to fray.

An architect by training, Wladimir analyzes the precarious state of the supply networks in the Samarkand microdistricts of the mid-1980s: "Because everything had been poorly made, the various components began to deteriorate very quickly." Leaky water pipes led to flooded basements; hot water for a bath or shower was a rarity. The district-heating system was in particularly bad shape: thermostats did not work, and valves were defective: "The heating system was dysfunctional practically everywhere." In addition to this hardship, residents experienced shortages of both the electricity and water, as neither the electrical network nor the water-supply system had been updated to accommodate increased demand.[74]

When confronted with such difficult circumstances, many people were compelled to return to the old practice of doing things themselves. The knowledge needed to intervene with modern technical systems was far greater than the skill required to repair a loam roof, however. When the water supply became erratic, Wladimir decided to take matters into his own hands—a project that required all of his "technical cunning." To collect water on the days when the supply was sufficient, Wladimir sourced "a 112-liter tank" and then "installed a water pump." The idea was to pump the water to the family's apartment on the fifth floor. As it turned out, the greatest challenge was to find suitable pipes. Whereas Wladimir was able

[73] The information in this paragraph is taken from an interview conducted by Mariya Petrova and Jonas van der Straeten on September 20, 2018, with Natalya (name changed).

[74] The quotations in this paragraph are taken from the interview mentioned in Footnote 46.

to buy valves and couplings in a local store, "pipes were in very short supply." So, again, Wladimir took the initiative. He told us:

> I went looking among the debris at various building sites, then I sawed them [the pipes], cut the threads myself, and made the connections. [...] Thank God our construction workers didn't really care; there were lots of scraps, and I designed the couplings myself.

Unfortunately, the situation became even worse after the fall of the USSR; for a long time, Uzbekistan was cut off from Tajikistan's power supply.[75] After independence, Wladimir claims, the situation "really turned into a nightmare. They couldn't keep the voltage stable, so we often had power outages. Household appliances and other machines suffered a lot." Natalya also maintains that service provision "collapsed" along with the Soviet Union.[76]

The "Private Initiative" for Home Improvement

Liberal ideology presumes that private ownership motivates people to take responsibility for their own property—for example, by carrying out repair-and-maintenance work. As early as the Cold War period, liberal Western scholars observed with satisfaction "the visible presence of private initiative" in Uzbekistan. Compared to other Soviet republics, Central Asia saw higher rates of home ownership. According to official figures from 1977, roughly half of all residential buildings—from freestanding houses to apartment complexes—in the region were privately owned, as compared to 13% in the Russian Soviet Federative Socialist Republic. Not surprisingly, US observers held a favorable view of this "benevolent attitude towards private housing" in the Central Asian Soviet Republics.[77]

Four of our interviewees grew up in adobe-brick courtyard houses, located in the older parts of Samarkand. In the Soviet era, while land was designated as State property, established residential areas were defined as

[75] Högselius, Per, "The Hidden Integration of Central Asia: The Making of a Region through Technical Infrastructures," *Central Asian Survey* 41 (2), 2022: 223–243: here: 235.

[76] The last quote comes from the interview with Natalya, mentioned in Footnote 73; all other quotes in this paragraph come from interviews with Wladimir (Footnote 46).

[77] The quotes and information in this paragraph are taken from Rywkin, Michael, "Housing in Central Asia: The Uzbek Example," in: Grant, Steven A., ed., *Soviet Housing and Urban Design*. Washington, DC: U.S. Department of Housing and Urban Development, 1980: 39–42; here: 40–41.

belonging to the so-called "private sector" (*chastnyj sektor*).[78] In practice, homeowners in both urban and rural areas of Samarkand had freedoms and responsibilities similar to homeowners in Western countries. To build additions onto homes or undertake other major renovations, permission from the authorities was officially required. According to Mukhabbat, however, almost "no one" applied for building permits during the Soviet era—or even bothered to register their houses with the land-registry office.[79] And, in the case of homeowners like Dilya's father, who cared little about repair and maintenance, the authorities would not intervene.

An interview with sister and brother Evgeniya and Boris corroborates these observations. The siblings were born in 1941 and 1939, respectively. They grew up in Samarkand's "colonial" district—the part of town planned and built by Russian military and administrative personnel in the last third of the nineteenth century. Evgeniya and Boris's grandfather, a Russian pharmacist, was dispatched to Russian Turkestan shortly before the First World War. Allotted a piece of land in the colonial city, he arranged to have a house built. To maximize insulation, the house was designed with one row of adobe and one row of burnt bricks with a layer of air in-between. In total, the wall was a staggering eighty centimeters thick. Around 1950, when the siblings were approximately ten years old, their parents decided to modernize parts of the building and undertake some repair work: the outdoor toilet was a mere hole in the ground, the kitchen needed an extension, and the cellar was threatening to collapse. Some of the planned work was quite complex; for example, the family needed to install heavy beams to buttress a new adobe wall. When asked if her parents had hired people to do the work, Evgeniya joked, "No, my father did everything, and for manpower, he used my mom." She laughed and pointed to herself: "I remember having to mix cement all the time." Boris assisted, as well: given that their father had lost one of his hands, Boris helped him to tile the terrace. So, more or less voluntarily, the family members did it themselves.[80]

Wladimir was not the only tenant in large-scale apartment houses who took matters into his own hands. A common practice across the Soviet

[78] Petrova, "*Nah am Boden*," p. 22.

[79] Interview with Mukhabbat mentioned in Footnote 56.

[80] The quotations and information in this paragraph are taken from an interview carried out by Mariya Petrova and Jonas van der Straeten on September 21, 2018, with Boris and Evgeniya (names changed).

Union and in other socialist countries was to convert balconies into bedrooms, kitchens, or storage areas. This practice became even more widespread in the post-Soviet era. Referring to a friend of his, a Khujand resident in 2010 told an interviewer, "There are seven people living in a two-room apartment, and he himself sleeps on the balcony."[81] Interviewees from the former Soviet republic of Georgia and the former Yugoslavia verify that enclosing balconies was a widespread practice during the Communist era.[82] Repurposing balconies helped to alleviate the general shortage of housing space; as such, authorities usually turned a blind eye to these unofficial renovations.

Natalya tells a similar story about Samarkand. After roughly two years of living in their home, her parents decided that the balcony was far too hot. Natalya notes, "We are in Central Asia; you cannot build an open balcony with metal balustrades," remarking that the design might work for Moscow, but not for Samarkand. To solve the problem, Natalya's father sourced bricks and window frames, then commissioned a couple of soldiers to do the work of enclosing the balcony. Despite the family's fear that the authorities would object, "nobody said anything." Forty-two years later, in 2018, the structure was still there.

As the supply of construction materials grew in the post-Soviet era, homeowners and tenants invested considerable resources in making their enclosed balconies more colorful and more elaborately designed. Balconies could even be made to reflect the dweller's personality. Referring to the bright-yellow color of his grandfather's balcony extension, one of Sgibnev's interviewees in 2010 comments, "This cheerful colour is there, because grandpa is a cheerful guy himself!"[83] To fully enclose the balcony, people used either glass, bricks, corrugated-iron sheets, or wood. More temporary enclosures were made of drywall or paperboard.[84] Turning a balcony into a kitchen required extending the water- and sewage-pipes. For residents who cook with coal, a chimney had to be added. Given that many residents undertook the works themselves, problems could ensue. A Khujand apartment owner tells the following story:

[81] Quoted from Sgibnev, "Remont: Housing," p. 186.

[82] Bouzarovski, Stefan, Joseph Saludvadze, and Michael Gentile, "A Socially Resilient Urban Transition? The Contested Landscapes of Apartment Building Extensions in Two Post-communist Cities," *Urban Studies* 48 (13), 2011: 2689–2714.

[83] Quoted from Sgibnev, "Remont: Housing," p. 190.

[84] Ibid., p. 188.

> My neighbour is flooding me regularly. Something is broken at her place and water seeps down to my balcony. Every time I go to her and tell her that she has to fix it, but she tells me: "I don't have any money". Her husband is somewhere, working, and she doesn't have sons. Well, I can't pressure her. I don't know. I live with it. I have done some repair work at my balcony, but it didn't help. [...] The problem is that she has transferred her kitchen to the balcony and apparently something was not done the right way. Now, water seeps through. Whatever I do, it doesn't work.[85]

This final quote highlights the neighbors' inability to solve the problem on their own. Both of them live in an apartment of their own, in a modernist apartment building, and neither one appears to have anyone to assist them; there is no social network to be mobilized. For reasons we can only speculate about, "We'll do it ourselves" is apparently not a viable option. Perhaps the problem is rooted in fragmented family ties, or the anonymity of life in a large apartment building. Or perhaps Anvar was right when he said: "Today, nobody needs anyone."

[85] Quoted from ibid., pp. 189–190. This quotation is taken *verbatim* from Sgibnev's English translation.

CHAPTER 10

Conclusion: Challenging Globalizing Technologies

> Scientific progress is a fraction, the most important fraction, of the process of intellectualization which we have been undergoing for thousands of years and which nowadays is usually judged in such an extremely negative way. Let us first clarify what this intellectualist rationalization, created by science and by scientifically oriented technology, means practically.
>
> Does it mean that we, today, for instance, everyone sitting in this hall, have a greater knowledge of the conditions of life under which we exist than has an American Indian or a Hottentot? Hardly. Unless he is a physicist, one who rides on the streetcar has no idea how the car happened to be set in motion. And he does not need to know. He is satisfied that he may "count" on the behavior of the streetcar, and he orients his conduct according to this expectation; but he knows nothing about what it takes to produce such a car so that it can move. The savage knows incomparably more about his tools. [...] The savage knows what he does in order to get his daily food and which institutions serve him in this pursuit. The increasing intellectualization and rationalization do *not*, therefore, indicate an increased and general knowledge of the conditions under which one lives.[1]

Here it is again, the infamous reference to "the savage." We first encountered this derogatory, racist denomination in the opening quote of

[1] Weber, Max, "Science as a Vocation," in: Gerth, H.H., and C. Wright Mills, eds, *From Max Weber: Essays in Sociology*. New York: Oxford University Press, 1958: 129–156, here: 138–139. Weber held this talk in 1917; it was first published in German in 1919.

M. Hård, *Microhistories of Technology*, Palgrave Studies in the History of Science and Technology,
https://doi.org/10.1007/978-3-031-22813-1_10

the introductory chapter, in which Helbig's use of "the savage" (*der Wilde*) referred to the Dayak gold miner. In the above excerpt, the slur connotes the "American Indian" or the "Hottentot," whom we today refer to as Native American and Khoekhoe people, respectively. The excerpt above comes from a well-known lecture delivered in 1917 by the German sociologist Max Weber. Like Helbig's observations on Dayak technology, Weber's remark about the knowledge and skills of indigenous people is based on a racist worldview. Paradoxically, both Helbig and Weber's repugnant comments were meant to demonstrate that "even" indigenous people, whom Europeans placed at the bottom of the cultural ladder, were neither incompetent nor ignorant. Despite their racist evolutionary thinking, Helbig and Weber acknowledged that Asian, African, and Native American peoples embody the skills and knowledge to command their resources—and their environment.

To my mind, Weber's reflections on the limitations of "rationalization" and of the development of "scientifically oriented technology" are of particular interest. Consistent with my argument in the preface of this book, Weber's ideas suggest that macro processes cannot be used to analyze the entire world. Even if a concept like globalization could be helpful in describing the rise of international connections, "globalization" fails to capture the particularities of local phenomena. The microhistories presented in this book were designed to do exactly that—to bring the local to the surface, as Ulf Hannerz suggested. The preceding chapters showcase selected technologies in various parts of the Global South and East. Incorporating primary-source research and oral testimonies, I have shown that the protagonists of these narratives created and maintained technological solutions that suited their needs. As the case of sugar manufacturers in Northern India indicates, it often makes sense to deem such solutions "appropriate technologies" (Chap. 4).

The case studies presented in this book span almost two hundred years. Many of the scenarios described took place in colonial settings. A central message is that historians of technology should not make the mistake of starting their analyses with the advent of colonialism. While colonial powers indeed brought with them their own tools and artifacts, they did not face a technological vacuum upon arrival. On the contrary, colonizers met with local material cultures that historians are compelled to acknowledge. When, for instance, European companies and colonial governments began to set up technological infrastructure, they did so in areas that had usually been inhabited for centuries. In the case of Côte d'Ivoire, before the

French colonizers began to design the first railroad lines, they made use of—and even expanded—the existing transportation network of footpaths and rivers. For centuries, this network had been an essential, well-integrated component of the "technological landscape" of the area (Chap. 3).

The process of widespread colonization certainly challenged established technological landscapes and material cultures. It would be inappropriate to interpret the technological exploitation of the colonies as a one-sided phenomenon, however. There is far more to the history of colonial technology than that which the catchphrase "tools of empire" claims.[2] As suggested by the microhistory about the introduction and use of electricity in East Africa, members of the non-European population quickly learned how to benefit from this new technology (Chap. 5). Customers and local technicians challenged the authority of so-called public utilities by developing intricate methods of accessing the electricity network. It is certainly no coincidence that this struggle between providers and users of electricity calls to mind what anthropologist Bryan Pfaffenberger dubbed "Technological Dramas."[3]

Pfaffenberger no doubt employed the concept of "drama" to draw attention to the conflicts that often attend the introduction of new technology. In doing so, Pfaffenberger also elevated the voices of users and consumers. To analyze user activities, he introduced concepts such as "counterappropriation" and "counterdelegation."

This book surpasses such concepts. Rather than focusing on how people in various parts of the Global South and East responded to the introduction of Western technologies, I have emphasized the persistence of established skills and methods. The availability of Western-imported products did not necessarily challenge the ways in which people went about their daily chores. For example, despite the availability of disposable sanitary pads—designed in the United States—in South Korea from the mid-1960s onward, a large majority of women continued to use the same kind of menstruation cloths their mothers and grandmothers had used (Chap. 8).

[2] Headrick, Daniel R., *The Tools of Empire: Technology and European Imperialism in the Nineteenth Century*. New York & Oxford: Oxford University Press, 1981.

[3] Pfaffenberger, Brian, "Technological Dramas," *Science, Technology, & Human Values* 17 (3), 1992: 282–312.

The microhistory of building methods in Central Asia illustrates a similar pattern of persistence (Chap. 9). Members of local communities preserved their knowledge of and skills with constructing, expanding, maintaining, and repairing their buildings made of friable adobe bricks. In labeling these collective construction and repair activities as part of a "Do-It-Yourself" (DIY) practice, I propose the global usefulness of this concept.

Aside from a steady persistence, the cultures described here were never static. Often, material cultures underwent change as an outcome of encounters with other cultures. For example, the concept of the "trading zone" describes sites at which representatives of various peoples and cultures exchanged skills, knowledge, and technologies (Chap. 2). Similarly, the "flexible settlement" encouraged change in the form of preparedness for innovation and an openness to novelty; in the case of Nairobi's low-income areas, technological change was an innate ability to adopt to constantly changing circumstances (Chap. 7).

Some people believe that, by narrating history "from below," the historian is automatically seduced into overlooking global connections. I have shown otherwise. For example, I have described the process of mining for gold in Borneo as an international endeavor traceable to the eighteenth century (Chap. 1). In line with this thinking, I have clearly identified the longstanding, transatlantic roots of creole cooking techniques in South America (Chap. 6). The 1890 cookbook *Eclectic Cooking* epitomizes how indigenous and European foodways synergize. Indeed, cross-cultural encounters often imply novelty. According to Hannerz, the concept of creole "has connotations of creativity and of richness of expression."[4] I couldn't agree more.

By "going global," the history of technology also becomes a more exciting endeavor and a richer academic discipline. I hope these microhistories will inspire readers to view the past in a new light. For example, this book makes clear that standard periodization paradigms are less useful than previously believed in our quest to accurately frame the past. In particular, the formal transition to a colonial power seldom coincides with the implementation of imperial tools on a broad scale (Chap. 3), just as the official status of "independent" does not mean independence from external engineering knowledge and foreign capital (Chap. 5). Another

[4] Hannerz, Ulf, *Transnational Connections: Culture, People, Places.* London: Routledge, 1996, p. 66.

example: although the First World War gave rise to fluctuations in the price of sugar on the "world market," these fluctuations barely affected the regional market in Northern India (Chap. 4).

Symmetry and Heterogeneity

The investigation of regions beyond Europe and North America increases the empirical basis of the history of technology. Recounting microhistories from around the world allows us to question received notions of world history, replete with its traditional myths and tenacious concepts. In fact, a paramount task of the humanities is to question established views and their attendant nomenclatures.

For example, in other writings, Jethron Ayumbah Akallah and I have shown the limitations of the Large-Technological-System (LTS) model of technological infrastructure.[5] The LTS is only one of several paradigms that a global history of technology is obliged to contest. The same applies to terms such as "technology transfer" and "technological diffusion." While acknowledging that material objects, knowledge, and skills often travel from one geographical area to another, I also suggest in this book that there are other stories to tell. The use of the terms "transfer" and "diffusion" suggest a benchmark view of technology. Such a view means that one specific technology or technological system serves as an exemplar for other solutions in the world. Benchmark technologies tend to spread and overtake other solutions. By default, then, the history of technology becomes one of growing uniformity and homogeneity. According to this narrative, people across the globe either willingly orient themselves around this exemplar—or are forced to do so for economic, political, or cultural reasons. In contrast, the microhistories in this book illustrate the heterogeneous nature of the world. The case histories are meant to support my argument that historians of technology would be well served to use the concept of "material culture" rather than "technology" when analyzing the complex connections between objects and societies.

Using the term material culture instead of technology provides several advantages. Increasingly, "technology" has come to be associated with complex, high-tech devices, as well as the marginalization of low-tech solutions. In fact, the term technology gives rise to a skewed perception of

[5] Akallah, Jethron Ayumbah, and Mikael Hård, "Under the Historian's Radar: Local Water Supply Practices in Nairobi, 1940–1980," *Water Alternatives* 13 (3), 2020: 886–901.

the material world: the association of "technology" with advanced machinery and systems can tempt historians of technology to overlook people that use manual tools and apply less "advanced" solutions. For example, whereas much has been written about the introduction of electricity-powered cooking appliances to modern households, considerably less has been written about low-tech food-preparation techniques. Taking "material culture" rather than "technology" as the point of departure, I demonstrated in Chap. 6 that barbequing meat on an outdoor grill was an integral part of Argentine material culture. Writing a history of material culture necessarily involves the analysis of manual techniques, local knowledge, and region-specific solutions.[6]

A further task of the humanities is to improve our understanding of human behavior and the complexities of human societies.[7] In my view, historical knowledge must aim at increasing our understanding of past events and cultures. Given that events and cultures are unique, this understanding requires knowledge about local conditions as well as empathy and an open mind. As the title of Benjamin Elman's history of Chinese science suggests, such understanding is only possible if we approach other cultures *On Their Own Terms*.[8] From a European standpoint, using a bird as a dowsing tool to locate gold may appear ineffectual, or even ridiculous. Similarly, scientifically educated agronomists would likely consider the use of low-yield, indigenous sugarcane varieties to be a strange—and even foolhardy—choice. The historian, however, is obliged to abjure verdicts, and must try to comprehend the beliefs and actions of others on their own terms.

In line with sociologist of science David Bloor, I suggest that understanding requires symmetry.[9] Symmetry implies that we—in an unbiased manner—must accept the fact that people in distinct settings and at various junctures make different choices. The fact that roughly half of women in Britain chose to adopt tampons in the 1960s and 1970s does not imply we can expect women in South Korea to have done the same. Symmetry also requires that historians—in a balanced manner—employ different

[6] El Hariry, Shorouk, et al., "Toward a Global History of Material Culture," *Technikgeschichte* 88 (2), 2021: 178–182.

[7] Wright, Georg Henrik von, *Explanation and Understanding*. London: Routledge & Kegan Paul, 1971.

[8] Elman, Benjamin A., *On Their Own Terms: Science in China, 1550–1900*. Cambridge, MA: Harvard University Press, 2005.

[9] Bloor, David, *Knowledge and Social Imagery*. 2nd ed. Chicago: University of Chicago Press, 1994 (orig. 1976).

theories, methods, concepts, and models for different cultures and for different eras. The choice of conceptual model must be culturally and historically sensitive. As such, we cannot expect the Large-Technological-System model, for example, to work for all regions of the Global South; we need other models, to address the circumstances in other parts of the world. At the same time, the historians of Europe must be open to theories, methods, concepts, and models that originate in other cultures. By adopting models more selectively, and by modifying favored approaches, we can ultimately develop commensurable paradigms.[10] To pick up on the "trading zone" concept of Chap. 2, I believe that one way to formulate cross-cultural interpretations could be to set up "trading zones" where historians from various parts of the world exchange views and experiences. This book is meant as a contribution to the trading zone of global historians of technology.

Clapperton Chakanetsa Mavhunga is one historian of technology who takes a truly non-Western perspective—and makes his material accessible to Western readers. Critiquing standard notions of technology and innovation for being biased in favor of Western-based concepts and knowledge systems, Mavhunga asks us to employ an analytic language that takes seriously "vernacular registers and syntaxes" from other parts of the world—in this case, Africa.[11] Taking a similar approach, scholars and coauthors John Law and Wen-yuan Lin write history from an East Asian perspective, carrying out "forms of postcolonial investigation that use non-Western analytical resources."[12] Further steps on the way toward a truly global history of technology and material culture require further forms of collaboration to foster a multilingual dialog and cross-cultural experience.

Globalization Revisited

We can now understand why "globalization" cannot serve as the central theme of a global history of technology in the nineteenth and twentieth centuries. Like Weber's "intellectualist rationalization," "globalization" is

[10] Kuhn, Thomas S., *The Structure of Scientific Revolutions.* Chicago: University of Chicago Press, 1962.

[11] Mavhunga, Clapperton Chakanetsa, "Introduction: What Do Science, Technology, and Innovation Mean from Africa?" in: idem, ed., *What Do Science, Technology, and Innovation Mean from Africa?* Cambridge, MA: MIT Press, 2017: 1–27, here: 7.

[12] Law, John, and Wen-yuan Lin, "Provincializing STS: Postcoloniality, Symmetry, and Method," *East Asian Science, Technology and Society* 11 (2), 2017: 211–227, here: 212.

an asymmetric—if not a teleological—concept. In this sense, globalization is on par with the concepts of "evolution," "transfer," and "diffusion." As long as we continue to believe that "mechanization" has taken command (Giedion) or that "Americanization" has been an "irresistible" force (de Grazia), we remain "one-dimensional" human beings (Marcuse).[13]

The world looks different from ground level. Everyday life is not characterized by smooth development processes ("evolution") but by mundane activities and conflict: we observe how the *chang'aa* distiller in Kibera struggled to contain production costs and to evade police, and we appreciate how, to this day, the home cook in South America makes elaborate dishes without having to invest in expensive new appliances. Indeed, daily life is not governed by one-way linearities, like "globalization," but by temporal rhythms.[14] For example, in Uzbekistan, a family decides to build onto their house only when the oldest son marries, and the first grandchildren are on the way. In India, agricultural workers harvested their sugarcane and produced their sugar in time to pay their taxes to the British colonialists, and in South Korea, women washed their menstruation pads in sync with their monthly periods. As global processes and larger networks operate in the background, local life continues.

History itself—microhistory, in particular—contains multitudes; the historian must make careful choices of what to narrate. Such choices reflect the historian's knowledge interests.[15] For this book, my knowledge interest has been to lay the foundation for a global history of technology that brings the local in contact with the global—both theoretically and empirically. Now, it is up to others to take the baton— and continue to craft further microhistories.

[13] Giedion, Sigfried. *Mechanization Takes Command: A Contribution to Anonymous History*. New York: Oxford University Press, 1948; Grazia, Victoria de, *Irresistible Empire: America's Advance through Twentieth-century Europe*. Cambridge, MA: Belknap Press, 2005; Marcuse, Herbert, *One-dimensional Man: Studies in the Ideology of Advanced Industrial Society*. Boston, MA: Beacon Press, 1964.

[14] Straeten, Jonas van der, and Heike Weber, "Technology and Its Temporalities: A Global Perspective," in: Carnino, Guillaume, et al., eds, *Global History of Techniques (19th–21st Centuries)*. Turnhout, Belgium: Brepols, forthcoming 2022.

[15] Habermas, Jürgen, *Technik und Wissenschaft als "Ideologie."* Frankfurt a.M.: Suhrkamp, 1968.

Bibliography

Archives

Archiv- und Museumsstiftung der Vereinten Evangelischen Mission, Wuppertal-Barmen, Germany

Archives nationales d'outre-mer, Aix-en-Provence, France

Archivo General de la Nación, Buenos Aires, Argentina

Basel Mission Archives/Mission 21, Basel, Switzerland

British Library, London, UK: India Office Records

Bundesarchiv, Berlin-Lichterfelde, Germany

National Archives, Kew, UK

Roemer- und Pelizaeus-Museum Hildesheim GmbH, Hildesheim, Germany

School of African and Oriental Studies (SOAS), University of London: Archives & Special Collections

Tanzania National Archives, Dar es Salaam, Tanzania

Interviews

Thirteen interviews (anonymized), carried out by Youngju Lee in Busan, Gangwon-do, Jeju, and Seoul, South Korea, Sept., 2017–April, 2018.

Ten interviews (anonymized), carried out by Mariya Petrova and Jonas van der Straeten, in Halle, Germany, June–July, 2017, and in Samarkand, Uzbekistan, Sept., 2018.

M. Hård, *Microhistories of Technology*, Palgrave Studies in the History of Science and Technology,
https://doi.org/10.1007/978-3-031-22813-1

Sources and Literature

Abashin, Sergey, "'Ideal'niy kolkhoz' v sovetskoi Sredney Azii: Istoriya neudachi ili uspekha?" *Acta Slavica Iaponica*, 29, 2011: 1–26.

Abbott, Elizabeth, *Sugar: A Bittersweet History.* London: Duckworth Overlook, 2009.

Adamovsky, Ezequiel, "Criollismo, experiencia popular y política: El gaucho como emblema subversivo," *Anuario del Instituto de Historia Argentina* 18 (1), 2018: e067, https://doi.org/10.24215/2314-257Xe06 [accessed Aug. 22, 2022].

Adas, Michael, *Machines as the Measure of Men: Science, Technology, and Ideologies of Western Dominance.* Ithaca, NY: Cornell University Press, 1989.

Ahn, Jae-Hee, "Analysis of Changes in Female Education in Korea from an Education-Labor Market Perspective," *Asian Women* 27 (2), 2011: 113–139.

Akala, Jethron Ayumba, *In the Technological Footprints of Urbanity: A Socio-political History of Water and Sanitation in Nairobi, 1899–2015.* Darmstadt: Technical University of Darmstadt, 2019. https://tuprints.ulb.tu-darmstadt.de/8550/ [accessed Aug. 19, 2022].

Akallah, Jethron Ayumbah, and Mikael Hård, "Under the Historian's Radar: Local Water Supply Practices in Nairobi, 1940–1980," *Water Alternatives* 13 (3), 2020: 886–901.

Alexander, Catherine, and Victor Buchli, "Introduction," in: Alexander, Catherine, Victor Buchli, and Caroline Humphrey, eds, *Urban Life in Post-Soviet Asia.* London: University College London Press, 2007: 1–39.

Allan, Keith, and Kate Burridge, *Forbidden Words: Taboo and the Censoring of Language.* Cambridge: Cambridge University Press, 2006.

Allworth, Edward, ed., *Central Asia: A Century of Russian Rule.* New York: Columbia University Press, 1967.

Amin, Shahid, *Sugar Cane and Sugar in Gorakhpur: An Inquiry into Peasant Production for Capitalist Enterprise in Colonial India.* Delhi: Oxford University Press, 1984.

Angerler, Hans, "Mission, Kolonialismus und Missionierte: Über die deutsche Batakmission in Sumatra," *Beiträge zur historischen Sozialkunde* 23 (2), 1993: 53–61.

Aran, Artemio, *¡Pampa!* Buenos Aires: Colección gaucha, 1943.

Archetti, Eduardo P., "Hibridación, pertenencia y localidad en la construcción de una cocina nacional," *Trabajo y sociedad* 2 (2), 2000, https://www.unse.edu.ar/trabajoysociedad/ARCHETTIFINAL.htm [accessed Aug. 22, 2022].

Arcondo, Aníbal, *Historia de la alimentación en Argentina: Desde los orígenes hasta 1920.* Córdoba, Argentina: Ferreyra, 2002.

"'Are Animals Technology?' Archive of Messages Posted on Envirotech," https://www.envirotechhistory.org/wp-content/uploads/2007/05/animaltech.pdf [accessed Aug. 19, 2022].

Aronson, Marc, and Marina Budhos, *Sugar Changed the World: A Story of Magic, Spice, Slavery, Freedom, and Science*. Boston, MA: Houghton Mifflin Harcourt, 2010.

Ashcroft, Bill, Gareth Griffiths, and Helen Tiffin, eds, *The Post-Colonial Studies Reader*, 2nd ed. London: Routledge, 2006.

Austin, Elisabeth, "Reading and Writing Juana Manuela Gorriti's *Cocina ecléctica*: Modeling Multiplicity in Nineteenth-Century Domestic Narrative," *Arizona Journal of Hispanic Cultural Studies* 12, 2008: 31–44.

Ayora-Diaz, Steffan Igor, "Home Kitchens: Techniques, Technologies, and the Transformation of Culinary Affectivity in Yucatán," in: idem., ed., *Cooking Technology: Transformations in Culinary Practice in Mexico and Latin America*. London: Bloomsbury, 2016: 85–98.

Baek, Min Jeong, "Wŏlgyŏnggyŏnghŏm-ŭl t'onghae pon yŏsŏng chŏngch'essŏng-ŭi hyŏngsŏng-kwa pyŏnhwa-Andong-si Pungsan-eup Sosan maŭr-ŭl chungsim-ŭro" (월경경험을 통해 본 여성 정체성의 형성과 변화-안동시 풍산읍 소산마을을 중심으로) ["Creation and Changes of Women's Identity According to Menstruation Experiences—Focusing on Sosan Village in Pungsan-eup Andong-si"], *MA Thesis*, Andong: Andong National University, 2012.

Baier, Andrea, et al., "Die Welt reparieren: Eine Kunst des Zusammenmachens," in: Baier, Andrea, et al., eds, *Die Welt reparieren: Open Source und Selbermachen als postkapitalistische Praxis*. Bielefeld: transcript, 2016: 34–62.

Balbi, Mariella, ed., *Los Chifas en el Perú: Historia y Recetas / Chinese restaurants, 'Chifas,' in Peru: History and Recipes*. Lima: Universidad San Martin de Porres, 1999.

Ballent, Anahi, "La 'casa para todos': Grandeza y miseria de la vivienda masiva," in: Devoto, Fernando, and Marta Madero, eds, *Historia de la vida privada en la Argentina: La Argentina entre multitudes y soledades*. Vol. 3: *De los años treinta a la actualidad*. Buenos Aires: Aguilar, Altea, Taurus, Alfaguara, 1999: 19–47.

Banerjee, Abhijit, and Esther Duflo, *Poor Economics: A Radical Rethinking of the Way to Fight Global Poverty*. New York: Public Affairs, 2011.

Banerjee, Himadri, *Agrarian Society of the Punjab (1849–1901)*. New Delhi: Manohar, 1982.

Barnes, J.H., *Progress Report for 1913 on Sugar in the Punjab*. Lahore: Punjab Government Press, 1913.

Barnes, J.H., *Sugarcane in the Gurdaspur District*. Lahore: Punjab Government Press, 1912.

Baudouin, Luc, and Patricia Lebrun, "Coconut (*Cocos nucifera* L.) DNA Studies Support the Hypothesis of an Ancient Austronesian Migration from Southeast Asia to America," *Genetic Resources and Crop Evolution* 56 (2), 2009: 257–262.

Bauer, Arnold J., *Goods, Power, History: Latin America's Material Culture*. Cambridge: Cambridge University Press, 2001.

Bauer, Kerstin, *Kleidung und Kleidungspraktiken im Norden der Côte d'Ivoire: Geschichte und Dynamiken des Wandels vom Ende des 19. Jahrhunderts bis zur Gegenwart.* Berlin: LIT, 2007.

Baviskar, B.S., "Indian Sugar Then and Now: Power to the Peasants," in: Moitt, Sugar, pp. 22–36.

Bayly, C.A., "The Origins of Swadeshi (Home Industry): Cloth and Indian Society, 1700–1930," in: Appadurai, Arjun, ed., *The Social Life of Things: Commodities in Cultural Perspective.* Cambridge: Cambridge University Press, 1986: 285–321.

Benson, C., *Sugar Mills*, Department of Land Records and Agriculture, Madras, Agricultural Branch: Bulletin No. 26. Madras: Government Press, 1892.

Berichte der Rheinischen Missions-Gesellschaft [*BRMG*], 1862–1912.

Bhabha, Homi K., *The Location of Culture.* London: Routledge, 1994.

Bianca, Stefano, *Hofhaus und Paradiesgarten: Architektur und Lebensform in der islamischen Welt.* Munich: C.H. Beck, 1991.

Binger, (Louis-Gustave), *Du Niger au Golfe de Guinée par le pays de Kong et le Mossi*, Vol. 1, Paris: Librairie Hachette, 1892.

Bloor, David, *Knowledge and Social Imagery.* 2nd ed. Chicago: University of Chicago Press, 1994 (orig. 1976).

Bodewes, Christine, "Chang'aa Drinking in Kibera Slum: The Harmful Effects of Contemporary Changes in the Production and Consumption of Traditional Spirits," *African Journal of Drug & Alcohol Studies* 9 (1), 2010: 49–55.

Bodewes, Christine, *Parish Transformation in Urban Slums: Voices of Kibera, Kenya.* Nairobi: Paulines Publications Africa, 2005.

Bonilla, Francisco A., *Pueblos jóvenes y urbanizaciones populares: Organización vecinal en Lima metropolitana y resto del Perú.* Lima: Ed. Mercurio, 1972.

Bonneuil, Christophe, "August Chevalier, savant colonial: Entre science et Empire, entre botanique et agronomie," in: Petitjean, Patrick, ed., *Les sciences colonials: Figures et institutions*, Vol. II. Paris: Orstom, 1996: 15–35.

Bosma, Ulbe, "Local Peasants and Global Commodities: Sugar Frontiers in India, Indonesia and the Philippines," *Österreichische Zeitschrift für Geschichtswissenschaften* 30 (3), 2019: 42–62.

Bouzarovski, Stefan, Joseph Saludvadze, and Michael Gentile, "A Socially Resilient Urban Transition? The Contested Landscapes of Apartment Building Extensions in Two Post-communist Cities," *Urban Studies* 48 (13), 2011: 2689–2714.

Bray, Francesca, "Flows and Matrices, Landscapes and Cultures," *ICON: Journal of the International Committee for the History of Technology* 22, 2016: 8–19.

Bray, Francesca, ed., *Rice: Global Networks and New Histories.* New York: Cambridge University Press, 2015.

Brennan, James R., and Andrew Burton, "The Emerging Metropolis: A History of Dar es Salaam, *circa* 1862–2000," in: Brennan, James R., Andrew Burton, and

Yusuf Lawi, eds, *Dar es Salaam: Histories from an Emerging African Metropolis.* Dar es Salaam: Mkuki na Nyota, 2007: 13–75.

Brusco, Josefina, *El amigo del hogar: Cocina criolla y extranjera con recetario para enfermos y convalescientes.* Lima: T. Scheuch, 1934.

Buchanan, Francis, *A Journey from Madras through the Countries of Mysore, Canara, and Malabar,* 3 Vols. London: Bulmer and Co., 1807.

Buchli, Victor, "Astana: Materiality and the City," in: Alexander, Catherine, Victor Buchli, and Caroline Humphrey, eds, *Urban Life in Post-Soviet Asia.* London: University College London Press, 2007: 40–69.

Buckley, Thomas, and Alma Gottlieb, "A Critical Appraisal of Theories of Menstrual Symbolism," in: idem., eds, *Blood Magic: An Anthropology of Menstruation.* Berkeley, CA: University of California Press, 1988: 3–50.

Burton, Andrew, "Townsmen in the Making: Social Engineering and Citizenship in Dar es Salaam, c.1945–1960," *The International Journal of African Historical Studies* 36 (2), 2003: 331–365.

Burton, Richard, and Nathaniel Ward, "Into Toba Batak Country, 1824," in: Reid, Anthony, ed., *Witnesses to Sumatra: A Travellers' Anthology.* Oxford: Oxford University Press, 1995: 175–192.

Caldo, Paula, *Mujeres cocineras: Hacia una historia sociocultural de la cocina Argentina a fines de siglo XIX y primera mitad del XX.* Rosario, Argentina: Prohistoria, 2009.

Cangaley, Manny, *The Voice of the Elders: An Evening of Storytelling with the Baoulé.* Cologne: Rüdiger Köppe, 2005.

Carrington, J.F., *Talking Drums of Africa.* New York: Negro Universities Press, 1969 (orig. 1949).

Chatriot, Alain, Marie-Emmanuelle Chessel, and Matthew Hilton, eds, *The Expert Consumer: Associations and Professionals in Consumer Society.* Aldershot: Ashgate, 2006.

Chevalier, Auguste, "Le Borassus æthiopum de l'Afrique Occidentale et son utilization," *Revue de botanique appliquée et d'agriculture coloniale* 10 (108), 1930: 649–655.

Chevalier, Auguste, "Nouvelles recherches sur les Palmiers du genre Raphia: (Suite et fin)," *Revue de botanique appliquée et d'agriculture coloniale* 12 (127), 1932: 198–213.

Cho, Haejoang, "Living with Conflicting Subjectivities: Mother, Motherly Wife, and Sexy Woman in the Transition from Colonial-modern to Postmodern Korea," in: Kendall, Laurel, ed., *Under Construction: The Gendering of Modernity, Class, and Consumption in the Republic of Korea.* Honolulu, HI: University of Hawai'i Press, 2002: 165–195.

Chung· kogyosaeng-ŭl wihan sŏ n'gyoyuk – kyehoek-kwa siltche (중.고교생을 위한 성 교육-계획과 실제) ["Sex Education for Middle and High-school Pupils: Plan and Practice"]. Seoul: Ewha Womans University, 1971.

Clozel, F.-J., *Dix ans à la Côte d'Ivoire*. Paris: Augustin Challamel, 1906.

Cocina peruana: Cocina China, dulces criollos, secretos de cocina. Lima: Bendezu, 1965.

Cohn, Benhard S., "Representing Authority in Victorian India," in: Hobsbawm and Ranger, Invention, pp. 165–209.

Cooke, Louise, "Earthen Building Materials and Techniques at Merv, Turkmenistan," in: *LEHM 2004: 4th International Conference on Building with Earth*. Weimar: Dachverband Lehm, 2004: 52–61.

Cowan, Ruth Schwartz, *More Work for Mother: The Ironies of Household Technology from the Open Hearth to the Microwave*. New York: Basic Books, 1983.

Cox, Francis Augustus, *History of the Baptist Missionary Society, from 1792 to 1842, Vol. I*. London: T. Ward and G. & J. Dyer, 1842.

Cuadra, Morena, and Morena Escardó, *The Big Peruvian Cookbook: 100 Delicious Traditional Recipes from Peru*. New York: Skyhorse Publishing, 2019.

Cuadra, Morena, and Morena Escardó, *The Peruvian Kitchen: Traditions, Ingredients, Tastes, and Techniques in 100 Delicious Recipes*. New York: Skyhorse, 2014.

Datta, R.L., and Tinkari Basu, "Improvements on the Country Process of Making Sugar from Gur," *Department of Industries, Government of Bengal, Bulletin No. 42*. Calcutta: Bengal Secretariat Book Depot, 1930.

Davis, Mike, *Planet of Slums*. London: Verso, 2006.

Deerr, Noël, in *The History of Sugar*, 2 Vols. London: Chapman and Hall, 1949–1950.

Delaney, Janice, Mary Jane Lupton, and Emily Toth, *The Curse: A Cultural History of Menstruation*. Urbana, IL: University of Illinois Press, 1988 (rev. ed.; orig. 1976).

Deutsches Kolonialblatt 17, 1906.

Deutsch-Ostafrikanische Zeitung, No. 82, 1908.

Dewall, H. von, "Aanteekeningen omtrent de noordoostkust van Borneo," *Tijdshrift voor indische Taal-, Land- en Volkenkunde* 4, 1855: 423–458.

DiMoia, John P., *Reconstructing Bodies: Biomedicine, Health, and Nation Building in South Korea since 1945*. Stanford, CA: Stanford University Press, 2013.

Di Nunzio, Marco, "'We are Good at Surviving': Street Hustling in Addis Ababa's Inner City," *Urban Forum* 23 (4), 2012: 433–447.

Dreyfus, Camille, *Six mois dans l'Attié (un Transvaal français)*. Paris: L. Henry May, 1900.

Dunn, P.D., *Appropriate Technology: Technology with a Human Face*. New York: Schocken, 1978.

East African Standard (Nairobi), Aug. 10, 1963; March 20, 1969a.

Edgerton, David, "Creole Technologies and Global Histories: Rethinking How Things Travel in Space and Time," *HoST: History of Science and Technology* 1, 2007: 75–112.

Edgerton, David, *The Shock of the Old: Technology and Global History since 1900*. London: Profile Books, 2006.

Edward, Frank, and Mikael Hård, "Maintaining the Local Empire: The Public Works Department of Dar es Salaam, 1920–1960," *The Journal of Transport History* 41 (1), 2020: 27–46.

El Hariry, Shorouk, et al., "Toward a Global History of Material Culture," *Technikgeschichte* 88 (2), 2021: 178–182.

Elman, Benjamin A., *On Their Own Terms: Science in China, 1550–1900*. Cambridge, MA: Harvard University Press, 2005.

Engel, Barbara, *Öffentliche Räume in den blauen Städten Russlands: Entwicklungen, Status und Perspektiven; funktionale und räumliche Anforderungen an die Erneuerung öffentlicher Räume in den neuen Industriestädten Sibiriens unter veränderten sozioökonomischen Bedingungen*. Tübingen: Wasmuth, 2004.

Erhard, Ludwig, *Wohlstand für alle*. Düsseldorf: Econ-Verlag, 1957.

Especialidades de la cocina criolla. Seleccionadas por el personal técnica de la revista Mucho Gusto. 2nd ed. Buenos Aires: Fabril, 1961 (orig. 1958).

Etherton, David, ed., *Mathare Valley: A Case Study of Uncontrolled Settlement in Nairobi*. Nairobi: Housing Research and Development Unit, University of Nairobi, 1971.

Faizulloev, Muhiddin, "Mahalla in Northern Tajikistan from the End of 19th to the Beginning of the 20th Century: The Case of Khujand and Its Suburbs," in: Roche, Sophie, ed., *The Family in Central Asia: New Perspectives*. Berlin: Klaus Schwarz Verlag, 2017: 55–83.

Farrell-Beck, Jane, and Laura Klosterman Kidd, "The Roles of Health Professionals in the Development and Dissemination of Women's Sanitary Products, 1880–1940," *The Journal of the History of Medicine and Allied Sciences* 51 (3), 1996: 325–352.

Fernández-Prieto, Leida, "Mapping the Global and Local Archipelago of Scientific Tropical Sugar: Agriculture, Knowledge, and Practice, 1790–1880," in: Manning, Patrick, and Daniel Rood, eds, *Global Scientific Practice in an Age of Revolutions, 1750–1850*. Pittsburgh, PA: University of Pittsburgh Press, 2016: 181–198.

Fernández-Souza, "Grinding and Cooking: An Approach to Mayan Culinary Technology," in: Ayora-Diaz, Steffan Igor, ed., *Cooking Technology: Transformations in Culinary Practice in Mexico and Latin America*. London: Bloomsbury, 2016: 15–27.

Fodde, Enrico, "Traditional Earthen Building Techniques in Central Asia," *International Journal of Architectural Heritage: Conservation, Analysis, and Restoration* 3 (2), 2009: 145–168.

Frederick, Christine, *Household Engineering: Scientific Management in the Home*. Chicago, IL: American School of Home Economics, 1920.

Freidenfelds, Lara, *The Modern Period: Menstruation in Twentieth-century America*. Baltimore, MD: Johns Hopkins University Press, 2009.

French, R.A., and F.E. Ian Hamilton, eds, *The Socialist City: Spatial Structure and Urban Policy*. Chichester: John Wiley, 1979.

Freytag, Mirjam, *Frauenmission in China: Die interkulturelle und pädagogische Bedeutung der Missionarinnen untersucht anhand ihrer Berichte von 1900 bis 1930*. Münster: Waxmann, 1994.

Fries, Eduard, *Rundbriefe*, 1, 1903 – 5, 1908. Berlin: Gutenberg.

Furth, Charlotte, and Ch'en Shu-yueh, "Chinese Medicine and the Anthropology of Menstruation in Contemporary Taiwan," *Medical Anthropology Quarterly* 6 (1), 1999: 27–48.

Galison, Peter, *Image and Logic: A Material Culture of Microphysics*. Chicago: The University of Chicago Press, 1997.

Galloway, J.H., *The Sugar Cane Industry: An Historical Geography from Its Origins to 1914*. Cambridge: Cambridge University Press, 1989.

Gandulfo, Petrona C. de, *El libro de Doña Petrona: Recetas de arte culinario*. Buenos Aires: no publisher, 1934.

Gandulfo, Petrona C. de, *Las recetas económicas de Doña Petrona*. Buenos Aires: Distal, 2002 (orig. 1962).

Gandy, Matthew, "Planning, Anti-planning and the Infrastructure Crisis Facing Metropolitan Lagos," *Urban Studies* 43 (2), 2006: 371–396.

Gangler, Anette, et al., eds. *Städte Usbekistans zwischen Tradition und Fortschritt: Städtische Transformationsprozesse der zentralasiatischen Städte Taschkent und Samarkand*. Cottbus and Stuttgart: BTU Cottbus and Stuttgart University, 2006.

Gelber, Steven M., "Do-It-Yourself: Constructing, Repairing and Maintaining Domestic Masculinity," *American Quarterly* 49 (1), 1997: 66–112.

Gerritsen, Anne, and Giorgio Riello, "Introduction: Writing Material Culture History," in: idem., eds, *Writing Material Culture History*. London and New York: Bloomsbury Academic, 2015: 1–13.

Ghanadan, Rebecca H., "Public Service or Commodity Goods? Electricity Reforms, Access, and the Politics of Development in Tanzania." *PhD Thesis*, University of California at Berkeley, 2008.

Ghobrial, John-Paul, ed., "Global History and Microhistory," *Past & Present*, Supplement 14, 2019a (special issue).

Ghobrial, John-Paul A., "Introduction: Seeing the World like a Microhistorian," *Past & Present*, Supplement 14, 2019b (special issue): 1–22.

Giedion, Sigfried. *Mechanization Takes Command: A Contribution to Anonymous History*. New York: Oxford University Press, 1948.

Ginzburg, Carlo, *The Cheese and the Worms: The Cosmos of a Sixteenth-century Miller*. Baltimore, MD: Johns Hopkins University Press, 1980 (orig. 1976).

Golubev, Alexey, and Olga Smolyak, “Making Things through Making Things: Soviet Do-It-Yourself Culture and Practices of Late Soviet Subjectivation,” *Cahiers du monde russe* 54 (3–4), 2013: 517–542.

Gorriti, Juana Manuela, *Cocina ecléctica* (edited by Camilla Cattarulla). Buenos Aires: La Crujia, 2014 (orig. 1890).

Gourlay, K.A., “Atumpan,” in: *Oxford Music Online*. Oxford University Press: https://doi.org/10.1093/gmo/9781561592630.article.L2232253 [accessed Aug. 7, 2020].

Grazia, Victoria de, “Changing Consumption Regimes in Europe, 1930–1970: Comparative Perspectives on the Distribution Problem,” in: Strasser, Susan, Charles McGovern, and Matthias Judt, eds, *Getting and Spending: European and American Consumer Societies in the Twentieth Century*. Cambridge: Cambridge University Press, 1998: 59–83.

Grazia, Victoria de, *Irresistible Empire: America's Advance through Twentieth-century Europe*. Cambridge, MA: Belknap Press, 2005.

Griggs, Peter D., *Global Industry, Local Innovation: The History of Cane Sugar Production in Australia, 1820–1995*. Berne, CH: Peter Lang, 2011.

Guardia, Sara Beatriz, “Gastronomía peruana del siglo XX: Historia e identidad,” in: Villavicencio, Maritza, ed., *Seminario historia de la cocina peruana*. Lima: Universidad de San Martin de Porres, 2007: 229–242.

Habermas, Jürgen, *Technik und Wissenschaft als “Ideologie.”* Frankfurt a.M.: Suhrkamp, 1968.

Hadi, Khan Bahadur S.M., *The Indian Sugar Industry*. Bhopal: Government of Bhopal, 1929.

Hadi, Saiyid Muhammad, *The Sugar Industry of the United Provinces of Agra and Oudh*. Allahabad: Government Press, 1902.

Hake, Andrew, *African Metropolis: Nairobi's Self-help City*. London: Sussex University Press, 1977.

Hannerz, Ulf, *Transnational Connections: Culture, People, Places*. London: Routledge, 1996.

Hård, Mikael, and Andrew Jamison, *Hubris and Hybrids: A Cultural History of Technology and Science*. Cambridge, MA: MIT Press, 2005.

Hård, Mikael, and Mai Lin Tjoa-Bonatz, “Trading Zones in a Colony: Transcultural Techniques at Missionary Stations in the Dutch East Indies, 1860–1940,” *Social Studies of Science* 50 (6), 2020: 932–955.

Hart, Keith, “Informal Income Opportunities and Urban Employment in Ghana,” *The Journal of Modern African Studies* 11 (1), 1973: 61–89.

HEA: Fachgemeinschaft für effiziente Energieanwendung e.V., ed., *Das blaue Kochbuch: Das Koch- und Backbuch für Anfänger und Fortgeschrittene*. 56th ed., Berlin: VDE Verlag, 2020.

Headrick, Daniel R., *The Tentacles of Progress: Technology Transfer in the Age of Imperialism, 1850–1940*. New York: Oxford University Press, 1988.

Headrick, Daniel R., *The Tools of Empire: Technology and European Imperialism in the Nineteenth Century*. New York: Oxford University Press, 1981.

Hege, Patrick, and Jonas van der Straeten, "Enclaves of Light and Citadels of Profit: Translating Electric Networks for Colonial Dar es Salaam, 1904–1910," *unpublished manuscript*, 2018.

Heinrich, Thomas, and Bob Batchelor, *Kotex, Kleenex, Huggies: Kimberly-Clark and the Consumer Revolution in American Business*. Columbus, OH: Ohio State University Press, 2004.

Helbig, Karl, *Eine Durchquerung der Insel Borneo (Kalimantan): Nach den Tagebüchern aus dem Jahre 1937*, Vol. 2. Berlin: Dietrich Reimer, 1982.

Helbig, Karl, "Gold bei den Dajaks," *Ostasiatische Rundschau* 19 (15/16), 1938: 386–390.

Herrle, Peter, Henning Lübbe, and Jakob Rösel, *Slums und Squatter-Siedlungen: Thesen zur Stadtentwicklung und Stadtplanung in der dritten Welt*. Stuttgart: University of Stuttgart, 1981.

Heßler, Martina, *"Mrs. Modern Woman": Zur Sozial- und Kulturgeschichte der Haushaltstechnisierung*. Frankfurt am Main: Campus, 2001.

Hill, Martin J. D., *The Harambee Movement in Kenya: Self-help, Development and Education among the Kamba of Kitui District*. London: The Athlone Press, 1991.

Hinostroza, Rodolfo, *Primicias de cocina Peruana*. Léon: Everest, 2006.

Hobsbawm, Eric, and Terence Ranger, eds, *The Invention of Tradition*. Cambridge: Cambridge University Press, 1983.

Hodge, Joseph Morgan, *Triumph of the Expert: Agrarian Doctrines of Development and the Legacies of British Colonialism*. Ohio University Press: Athens, OH, 2007.

Högselius, Per, "The Hidden Integration of Central Asia: The Making of a Region through Technical Infrastructures," *Central Asian Survey* 41 (2), 2022: 223–243.

Houdaille, M., *Le chemin de fer et le port de la Côte d'Ivoire: Organisation d'une entreprise coloniale*. Paris and Nancy: Berger-Levrault, 1905.

Hughes, Thomas P., "The Evolution of Large Technological Systems," in: Bijker, Wiebe E., Thomas P. Hughes, and Trevor J. Pinch, eds, *The Social Construction of Technological Systems: New Direction in the Sociology and History of Technology*. Cambridge, MA: MIT Press, 1987: 51–82.

Hughes, Thomas P., *Networks of Power: Electrification in Western Society, 1880–1930*. Baltimore, MD: Johns Hopkins University Press, 1983.

Humburg, Martin, Dominik Bonatz, and Claus Veltmann, eds, *Im "Land der Menschen": Der Missionar und Maler Eduard Fries und die Insel Nias*. Bielefeld: Regio, 2003.

Humle, William, and R. P. Sanghi, *The Improvement of the Indigenous Methods of* Gur *and Sugar Making in the United Provinces*, Bulletin No. 82 of the

Agricultural Research Institute, Pusa. Kolkata: Superintendent Government Printing, 1918.

Humle, William, and R. P. Sanghi, *A Note on the Improvement of the Indigenous Methods of Gur and Sugar Making in the United Provinces: A Report on the Government Experimental Sugar Factory, Nawabganj, Bareilly District.* Allahabad: Government Press, United Provinces, 1916.

Jackson, James C., *Chinese in the West Borneo Goldfields: A study in Cultural Geography*. Hull: University of Hull Publications, 1970.

Jajeśniak-Quast, Dagmara, "In the Shadow of the Factory: Steel Towns in Postwar Eastern Europe," in: Hård, Mikael, and Thomas J. Misa, eds, *Urban Machinery: Inside Modern European Cities*. Cambridge, MA: MIT Press, 2008: 187–210.

Jeon, Gyeong-ok, et al., *Hangug-yeoseong-inmulsa1 - hangug-yeoseong-geunhyeondaesa1. hangaehwagi~1945nyeon* (한국여성인물사1 - 한국여성근현대사1. 한개화기~1945년) ["The History of Korean Female Figures 1: The Modern and Contemporary History of Korean Women 1: The Han Period till ca. 1945"], Seoul: Asian Women's Research Institute, Sookmyung Women's University, 2004.

Johansson, Birgitta, *Ny teknik och gamla vanor: En studie av mikrovågsugnens introduktion*. Linköping: Linköping University, 1988.

Jordan, Jens W., "Die Mahallas der timuridischen Altstadt Samarkands," in: *2750 Jahre Samarkand*. Erfurt: Stein, 2007: 58–65.

Joseph, Gaston, "Exploitation indigène de l'or en Côte-d'Ivoire," *Bulletins et mémoires de la Société d'anthropologie de Paris*, 6th series, 4 (3–4), 1913: 372–375.

Josephson, Paul R., *New Atlantis Revisited: Akademgorodok, the Siberian City of Science*. Princeton, NJ: Princeton University Press, 1997.

Josephson, Paul R., "'Projects of the Century' in Soviet History: Large-scale Technologies from Lenin to Gorbachev," *Technology and Culture* 36 (3), 1995: 519–559.

Kim, Che-han, "Han'gukyŏsŏng-ŭi ch'ogyŏng-e kwanhan chosayŏn'gu" (韓國女性의 初經에 관한 調査硏究) [authorized title: "A Survey on Woman's Menarchal in Korea"], *Nonmunchip* (論文集) ["Collection of Treatises, Seoul National University of Education"] 3, 1970: 167–180.

Kim, Deok-ho, "Han'guk-esŏŭi ilsangsaenghwal-kwa sobi-ŭi migukhwa munje" (한국에서의 일상 생활과 소비의 미국화 문제) ["Issues of Americanization in Everyday Life and Consumption in Korea"], in: Kim, Deok-ho, and Won Yong-jin, eds, *Amerik'anaijeisyŏn* (아메리카나이제이션) ["Americanization"], Seoul: P'urŭnyŏksa 2008: 121–158.

Kim, Soo-Jin, "Yŏsŏngŭibok-ŭi pyŏnch'ŏn-ŭl t'onghae pon chŏnt'ong-kwa kŭndae-ŭi chendŏjŏngch'i - haebang ihu ~ 1960 nyŏndae ch'oban-ŭl chungsim-ŭro" (여성의복의 변천을 통해 본 전통과 근대의 젠더정치-해방 이후~1960년대 초반을 중심으로) [authorized translation: "Reading the Transformations of

Women's Clothes in Postcolonial Korea: Gender Politics of Tradition and Modernity"], *P'eminijŭm myŏn'gu* (페미니즘연구) ["Issues in Feminism"] 7 (2), 2007: 281–320.

Kim, Soojung, "Tarŭn sesang : Kaejim, taean saengnidae iyagi" (다른 세상: 개짐, 대안 생리대 이야기) ["Alternative World: A Story of an Alternative Sanitary Napkin"], *Yŏ/songiron* (여/성이론) ["Journal of Feminist and Sexual Theories"] 10 (7), 2004: 209–215.

Klein, Dieter, ed., Jehova se nami nami*: Die Tagebücher der Johanna Diehl, Missionarin in Deutsch-Neuguinea 1907–1913.* Wiesbaden: Harrassowitz Verlag, 2005.

Konrad, Dagmar, *Missionsbräute: Pietistinnen des 19. Jahrhunderts in der Basler Mission.* Münster: Waxmann, 2001.

Koroteyeva, Victoria, and Ekaterina Makarova, "Money and Social Connections in the Soviet and Post-Soviet Uzbek City," *Central Asian Survey* 17 (4), 1998: 579–596.

Kotkin, Stephen, *Magnetic Mountain: Stalinism as a Civilization.* Berkeley, CA: University of California Press, 1995.

Kuhn, Thomas S., *The Structure of Scientific Revolutions.* Chicago: University of Chicago Press, 1962.

Kumar, Ashutosh, *Coolies of the Empire: Indentured Indians in the Sugar Colonies, 1830–1920.* Cambridge: Cambridge University Press, 2017.

Kyunghyang Shinmun (경향신문) ["Kyunghyang Newspaper"] (Seoul), April 5, 1966; March 10, 1975; Aug. 11, 1977.

Laborde, Gustavo, *El asado: Origen, historia, ritual.* Montevideo: Ediciones de la Banda Oriental, 2010.

Landström, Catharina, "National Strategies: The Gendered Appropriation of Household Technology," in: Hård, Mikael, and Andrew Jamison, eds, *The Intellectual Appropriation of Technology: Discourses on Modernity, 1900–1939.* Cambridge, MA: MIT Press, 1998: 163–188.

Larkin, Brian, *Signal and Noise: Media, Infrastructure, and Urban Culture in Nigeria.* Durham, NC, & London: Duke University Press, 2008.

Laszczkowski, Mateusz, "Scraps, Neighbors, and Committees: Material Things, Place-Making, and the State in an Astana Apartment Block," *City & Society* 27 (2), 2015: 136–159.

Law, John, and Wen-yuan Lin, "Provincializing STS: Postcoloniality, Symmetry, and Method," *East Asian Science, Technology and Society* 11 (2), 2017: 211–227.

Law, Robin, "Horses, Firearms, and Political Power in Pre-colonial West Africa," *Past and Present* 72 (1), 1976: 112–132.

Law, Robin, *The Slave Coast of West Africa, 1550–1750: The Impact of the Atlantic Slave Trade on an African Society.* Oxford: Clarendon, 1991.

Lecciones de cocina: Contienen las más deliciosas y fáciles recetas de cocina como también consejos prácticos para el hogar. Lima: Imprenta del estado, 2nd ed., 1921.

Lee, Gilsang, et al., eds, *The History of Education in Korea: A Sourcebook*. Seongnam: The Academy of Korean Studies Press, 2015.

Lee, Hwa-hyung, et al., *Han'gukhyŏndaeyŏsŏng-ŭi ilsangmunhwa 8. Kajŏngwisaeng* (이화형 외, 한국현대여성의 일상문화 8. 가정위생) ["The Daily Culture of Modern Korean Women, 8: Home Hygiene"], Seoul: Kuk'akcharyowŏn, 2005.

Lee, Sang-rok, "1970 nyŏndae sobiŏkchejŏngch'aek-kwa sobimunhwa-ŭi ilsangjŏngch'ihak" (1970년대 소비억제정책과 소비문화의 일상정치학) [authorized translation: "The Consumption Control Policy and the Politics of Everyday Life of Consumer Culture in the 1970s"], *Yŏksamunjeyŏn'gu* (역사문제연구) ["Critical Studies on Modern Korean History"] 17 (1), 2013: 137–182.

Lee, Won-gu, *Sojung-han uri-dŭl - 5 · 6 hangnyŏn* (소중한 우리들 – 5 · 6 학년) ["Precious Us: 5th and 6th Grade"]. Seoul: Tongjinmunhwasa, 1984.

Lee, Young Ju, Irhoeyong saengnidae-ŭi toip-kwa wŏlgyŏng-hanŭn mom-e tae-han insik-ŭi pyŏnhwa – 1960-1980 nyŏndae han'guk-ŭi kyŏngu (일회용 생리대의 도입과 월경하는 몸에 대한 인식의 변화 - 1960-1980 년대 한국의 경우) [authorized translation: "The Introduction of Disposable Menstrual Products and the Change of Perception about the Menstrual Body, in 1960s–1980s Korea"], *M.Sc. Thesis*, Seoul: Seoul National University, 2018, https://dcollection.snu.ac.kr/srch/srchDetail/000000152465.

Leyden, John, *Historical Account of Discoveries and Travels in Africa*, Vol. II. Edinburgh: George Ramsay, 1817.

Lie, Merete, and Knut Sørensen, eds, *Making Technology Our Own? Domesticating Technology into Everyday Life*. Oslo: Scandinavian University Press, 1996.

Lin, Shing-ting, "'Scientific' Menstruation: The Popularisation and Commodification of Female Hygiene in Republican China, 1910s–1930s," *Gender & History* 25 (2), 2013: 294–316.

Lindell, Ilda, "Introduction: The Changing Politics of Informality—Collective Organizing, Alliances and Scales of Engagement," in: idem., ed., *Africa's Informal Workers: Collective Agency, Alliances and Transnational Organizing in Urban Africa*. London: Zed Books, 2010: 1–30.

Lindell, Ilda, and Mats Utas, "Networked City Life in Africa: Introduction," *Urban Forum* 23 (4), 2012: 409–414.

Lindqvist, Svante, "Changes in the Technological Landscape: The Temporal Dimension in the Growth and Decline of Large Technological Systems," in: idem., *Changes in the Technological Landscape: Essays in the History of Science and Technology*. Sagamore Beach, MA: Science History Publications, 2010: 3–24.

Long, E. Rodriguez, and Jewel B. Groves, *El asado criollo—Roast Spit Barbeque*. Buenos Aires: Bell, 1963

Long, Pamela O., "Trading Zones in Early Modern Europe," *Isis* 106 (4), 2015: 840–847.

Lorentzen, Håkon, and Line Dugstad, *Den norske dugnaden: Historie, kultur og fellesskap*. Kristiansand: Norwegian Academic Press, 2011.

Lynn, Martin, *Commerce and Economic Change in West Africa: The Palm Oil Trade in the Nineteenth Century*. Cambridge: Cambridge University Press, 1997.

Magaña González, Claudia Rocío, "Technologies and Techniques in Rural Oaxaca's Zapotec Kitchens," in: Ayora-Diaz, Steffan Igor, ed., *Cooking Technology: Transformations in Culinary Practice in Mexico and Latin America*. London: Bloomsbury, 2016: 55–67.

Mahmud, Shihabuddin, and Umit Duyar-Kienast, "Spontaneous Settlements in Turkey and Bangladesh: Preconditions of Emergence and Environmental Quality of Gecekondu Settlements and Bustees," *Cities* 18 (4), 2001: 271–280.

Maines, Rachel P., *Hedonizing Technologies: Paths to Pleasure in Hobbies and Leisure*. Baltimore, MD: Johns Hopkins University Press, 2009.

Majumdar, Rochona, *Writing Postcolonial History*. London: Bloomsbury, 2010.

Marcuse, Herbert, *One-dimensional Man: Studies in the Ideology of Advanced Industrial Society*. London: Routledge & Kegan Paul, 1964.

Marschall, Wolfgang, *Der Berg des Herrn der Erde: Alte Ordnung und Kulturkonflikte in einem indonesischen Dorf*. Munich: Deutscher Taschenbuch Verlag, 1976.

Marta, *La cocinera criolla y recetario curativo doméstico*. 18th ed. Barcelona: Luis Gili, 1928.

Marteau D'Autry, Christilla, "*Vyjdem vse, kak odin!* 'Allons-y tous comme un seul homme!' Ethnographie d'un *hashar* national dans un quartier de Samarkand, Ouzbékistan," *Cahiers d'Asie centrale* 19–20, 2011: 279–301.

The Martyr of Sumatra: A Memoir of Henry Lyman. New York: Robert Carter & Brothers, 1856.

Mavhunga, Clapperton Chakanetsa, "Introduction: What Do Science, Technology, and Innovation Mean from Africa?" in: idem, ed., *What Do Science, Technology, and Innovation Mean from Africa?* Cambridge, MA: MIT Press, 2017: 1–27.

Mayne, Alan, *Slums: The History of a Global Injustice*. London: Reaktion Books, 2017.

Mchome, Emanuel Lukio, *"Blackout Blues": A Socio-cultural History of Vulnerable Electricity Networks and Resilient Users in Dar es Salaam, 1920–2020*, Darmstadt: Technical University of Darmstadt, 2022. https://tuprints.ulb.tu-darmstadt.de/id/eprint/21803 [accessed Aug. 30, 2022].

McShane, Clay, and Joel A. Tarr, *The Horse in the City: Living Machines in the Nineteenth Century*. Baltimore, MD: Johns Hopkins University Press, 2007.

McVicar, Kenneth Gordon, "Twilight of an East African Slum: Pumwani and the Evolution of African Settlement in Nairobi," *PhD Dissertation*, University of California Los Angeles, 1968.

Medick, Hans, *Weben und Überleben in Laichingen 1650–1900: Lokalgeschichte als Allgemeine Geschichte*. Göttingen: Vandenhoeck & Ruprecht, 1996.

Meerovich, Mark G., Viachaslau I. Menkouski, and Igor L. Zherebtsov, *"Socialist City": Idea and Its Realization in the Soviet Union 1920's and 1930's*. Belianum: Banská Bystica, Slovakia, 2019.

Menzel, Gustav, *Die Rheinische Mission: Aus 150 Jahren Missionsgeschichte*. Wuppertal: Verlag der Vereinigten Evangelischen Mission, 1978.

Merskin, Debra, "Adolescence, Advertising, and the Ideology of Menstruation," *Sex Roles* 40 (11–12), 1999: 941–957.

Mintz, Sidney W., *Sweetness and Power: The Place of Sugar in Modern History*. New York: Penguin, 1985.

Misa, Thomas J., *Leonardo to the Internet: Technology and Culture from the Renaissance to the Present*. Baltimore, MD: Johns Hopkins University Press, 2004.

Moitt, Bernard, ed., *Sugar, Slavery, and Society: Perspectives on the Caribbean, Indian, the Mascarenes, and the United States*. Gainesville, FL: University Press of Florida, 2004.

Molohan, M.J.B., *Detribalization: A Study of the Areas of Tanganyika where Detribalized Persons are Living*. Dar es Salaam: Government Press, 1959.

Monnier, Marcel, *France noire (Côte D'Ivoire et Soudan)*, Paris: Librairie Plon, 1894.

Moon, Kook-Hyun et al., *Segye-ka paeunŭn han'guk kiŏp-ŭi hŭimang YuhanKimberly* (세계가 배우는 한국 기업의 희망 유한킴벌리) ["A Role Model for Korean companies: Yuhan-Kimberly that the World is Following"]. Seoul: Hans Media, 2005.

Moon, Seungsook, *Militarized Modernity and Gendered Citizenship in South Korea*. Durham, NC: Duke University Press, 2005.

Moon, Suzanne, *Technology and Ethical Idealism: A History of Development in the Netherlands East Indies*. Leiden: CNWS Publications, 2007.

Moreland, W.H., "Introduction," in: Hadi, S.M., *Improvements in Native Methods of Sugar Manufacture*. Agricultural Series, Bulletin No. 19, Department of Land Records and Agriculture, United Provinces of Agra and Oudh. Allahabad: Government Press, 1905: 1–3.

Möser, Kurt, "Thesen zum Pflegen und Reparieren in den Automobilkulturen am Beispiel der DDR," *Technikgeschichte* 79 (3), 2012, 207–226.

Müller, Simone M., *Wiring the World: The Social and Cultural Creation of Global Telegraph Networks*. New York: Columbia University Press, 2016.

Nelson, Laura, "South Korean Consumer Nationalism: Women, Children, Credit, and Other Perils," in: Garon, Sheldon, and Patricia L. Maclachlan, eds, *The Ambivalent Consumer: Questioning Consumption in East Asia and the West*. Ithaca, NY: Cornell University Press, 2006: 188–207.

Nye, David E., *American Technological Sublime*. Cambridge, MA: MIT Press, 1994.

Nye, David E., *Electrifying America: Social Meanings of a New Technology, 1880–1940*. Cambridge, MA: MIT Press, 1990.

Nye, David E., *Narratives and Spaces: Technology and the Construction of American Culture*. Exeter: University of Exeter Press, 1998.

Ogunremi, Gabriel Ogundeji, *Counting the Camels: The Economics of Transportation in Pre-industrial Nigeria*. New York: NOK Publishers, 1982.

Oldenziel, Ruth, and Karin Zachmann, eds, *Cold War Kitchen: Americanization, Technology, and European Users*. Cambridge, MA: MIT Press, 2009.

Oldenziel, Ruth, and Mikael Hård, *Consumers, Tinkerers, Rebels: The People Who Shaped Europe*. Houndmills, Basingstoke: Palgrave Macmillan, 2013.

Osterhammel, Jürgen, *The Transformation of the World*. Princeton: Princeton University Press, 2014 (orig. 2009).

Oudshoorn, Nelly, and Trevor J. Pinch, eds, *How Users Matter: The Co-construction of Users and Technology*. Cambridge, MA: MIT Press, 2003.

Para ti (Buenos Aires), Vol. 39, 1960.

Pa´til, P.C., *Poona Furnace*, Department of Agriculture, Bombay: Bulletin No. 48. Bombay: Government Central Press, 1911.

Pérez, Inés, "Corazón de hojalata, hogar de terciopelo: La cocina, epicentro del mundo doméstico (Mar del Plata – Argentina, 1950–1970)," *Asparkía: Investigació feminista* 21, 2010: 105–128.

Perks, Robert, and Alistair Thomson, *The Oral History Reader*, 3rd ed. London: Routledge, 2016.

Perrin, Noel, *Giving Up the Gun: Japan's Reversion to the Sword, 1543–1879*. Boston, MA: David R. Godine, 1979.

Petersen, Sonja, "*Das elektrische Kochen* – Die Vollelektrische Küche als Leitbild moderner Haushaltsführung," *Food & History* 11, 2013: 75–106.

Pétric, Boris Mathieu, "The Mahalla in Present-day Uzbekistan," in: Gammer, Moshe, ed., *The Caspian Region, Volume I: A Re-emerging Region*. London: Routledge, 2004: 228–243.

Petrova, Mariya, *"Nah am Boden": Privater Hausbau zwischen Wohnungsnot und Landkonflikt im sowjetischen Samarkand der 50er und 60er Jahre*. Berlin: De Gruyter, 2020.

Pfaffenberger, Brian, "Technological Dramas," *Science, Technology, & Human Values* 17 (3), 1992: 282–312.

Philips, John Edward, ed., *Writing African History*. Rochester, NY: University of Rochester Press, 2005.

Pilcher, Jeffrey M., "Eating à la Criolla: Global and Local Foods in Argentina, Cuba, and Mexico," *IdeAs: Idées d'Amériques* 3, 2012, https://doi.org/10.4000/ideas.406 [accessed Aug. 22, 2022].

Pino, Julio César, "Sources on the History of Favelas in Rio de Janeiro," *Latin American Research Review* 32 (3), 1997: 111–122.

Pisarčik, A.K., *Narodnaya arkhitektura Samarkanda: XIX–XX vv.* Dushanbe: Izdatel'stvo "Donish," 1974.

Pite, Rebekah E., "La cocina criolla: A History of Food and Race in Twentieth-century Argentina," in: Alberto, Paulina L., and Eduardo Elena, eds, *Rethinking Race in Modern Argentina.* New York: Cambridge University Press, 2016: 99–125.

Pite, Rebekah E., *Creating a Common Table in Twentieth-century Argentina: Doña Petrona, Women, and Food.* Chapel Hill: University of North Carolina Press, 2013.

Portères, Roland, "Le palmier ronier (*Borassus aethiopum* Mart.) dans la Province du Baoule (Côte d'Ivoire)", *Journal d'agriculture tropicale et de botanique appliquée* 11 (12), 1964: 499–514.

Posewitz, Theodor, *Borneo: Entdeckungsreisen und Untersuchungen. Gegenwärtiger Stand der geologischen Kenntnisse. Verbreitung der nutzbaren Mineralien.* Berlin: R. Friedländer & Sohn, 1889.

Posewitz, Theodor, "Das Goldvorkommen in Borneo," *Mittheilungen aus dem Jahrbuche der kön. ungarischen geologischen Anstalt* 6, 1883: 175–190.

Prestholdt, Jeremy, *Domesticating the World: African Consumerism and the Genealogies of Globalization.* Berkeley: The University of California Press, 2008.

Ray, Utsa, *Culinary Culture in Colonial India: A Cosmopolitan Platter and the Middle-Class.* Cambridge: Cambridge University Press, 2015.

Recetas selectas para cocinas modernas. Lima, Peru: Empresas Eléctricas Asociadas, 1936.

Report of the Indian Sugar Committee. Simla: Government Central Press, 1921.

Richards, P.B. "Foreword," in: Hadi, Khan Bahadur Syed Mohammad, *Rab, Khand, and Gur Making with Hadi Miniature Bel.* Bulletin No. 74, Department of Agriculture, United Provinces. Allahabad: Superintendent, 1938: i.

Riedijk, W., ed., *Appropriate Technology for Developing Countries.* Delft: Delft University Press, 1982.

Robinson, Charles Henry, *Hausaland, or Fifteen Hundred Miles through the Central Soudan.* London: Sampson Low, Marston and Co., 1896.

Roh, Jieun, "1970 nyŏndae~1990 nyŏndae saengnidae kwanggo tamnon-kwa yŏsŏng (1970년대~1990년대 생리대 광고 담론과 여성)" [authorized translation: "1970s~1990s Menstrual Product Advertisements and Gender in Korea"], *Yŏsŏng-kwa yŏksa* (여성과 역사) ["Women and History"] 21, 2014: 219–49.

Roh, Jieun, "Wŏlgyŏng kyŏnghŏm-kwa munhawjŭk kŭmgi" (월경 경험과 문화적 금기) [authorized translation: "A Study of the Experiences of Menstruation and the Cultural Taboos"], *MA Thesis*, Seoul: Ewha Womans University, 1995.

Romaine, Suzanne, *Pidgin and Creole Languages.* London: Longman, 1988.

Roszak, Theodore, "Introduction," in: Schumacher, Small, pp. 1–9.

Rywkin, Michael, "Housing in Central Asia: The Uzbek Example," in: Grant, Steven A., ed., *Soviet Housing and Urban Design*. Washington, DC: U.S. Department of Housing and Urban Development, 1980: 39–42.

Russell, Andrew L., and Lee Vinsel, "After Innovation, Turn to Maintenance," *Technology and Culture* 59 (1), 2018, 1–25.

Said, Edward, *Orientalism*. New York: Pantheon, 1978.

Sardar, Ziauddin, *Postmodernism and the Other: The New Imperialism of Western Culture*. London: Pluto Press, 1998.

Şaul, Mahir, "The War Houses of the Watara in West Africa," *The International Journal of African Historical Studies* 31 (3), 1998: 537–570.

Schatzberg, Eric, *Technology: Critical History of a Concept*. Chicago, IL: University of Chicago Press, 2018.

Schivelbusch, Wolfgang, *The Railway Journey: Trains and Travel in the 19th Century*. New York: Urizen Books, 1979 (orig. 1977).

Schott, Dieter, *Die Vernetzung der Stadt: Kommunale Energiepolitik, öffentlicher Nahverkehr und die 'Produktion' der modernen Stadt. Darmstadt—Mannheim—Mainz 1880–1918*. Darmstadt: Wissenschaftliche Buchgesellschaft, 1997.

Schulze, Winfried, ed., *Sozialgeschichte, Alltagsgeschichte, Mikro-Historie: Eine Diskussion*. Göttingen: Vandenhoeck & Ruprecht, 1994.

Schumacher, E.F., *Small is Beautiful: A Study of Economics as if People Mattered*. London: Abacus, 1973.

Schwartz, Stuart B., ed., *Tropical Babylons: Sugar and the Making of the Atlantic World, 1450–1680*. Chapel Hill, NC: University of North Carolina Press, 2004.

Scott, James C., *Seeing Like a State: How Certain Schemes to Improve the Human Condition Have Failed*. New Haven, CT: Yale University Press, 1998.

Seeley, Ivor H., *Building Maintenance*. London: Macmillan, 1976.

Seok-jin Jang, "Urinara yŏsŏng saengniyongp'um chejoŏp-ŭi maak'et'ing-e kwanhan yŏn'gu" (우리나라 女性 生理用品 製造業의 마아케팅에 관한 硏究) ["A Study on the Marketing Female Menstrual Products in Korea"]. *Master's Thesis*, Korea University School of Management, 1981.

Sgibnev, Wladimir, "Remont: Housing Adaptation as Meaningful Practice of Space Production in Post-Soviet Tajikistan", *Europa Regional* 22 (1–2), 2014: 53–64.

Sgibnev, Wladimir, "*Remont*: The Social Production of Space in Central Asia," *PhD Dissertation*. Berlin: Humboldt University, 2012.

Shove, Elizabeth, *Comfort, Cleanliness and Convenience: The Social Organization of Normality*. Oxford: Berg, 2003.

Simone, AbdouMaliq, "People as Infrastructure: Intersecting Fragments in Johannesburg," *Public Culture* 16 (3), 2004: 407–429.

Smith, Andrew F., *Sugar: A Global History*. London: Reaktion Books, 2015.

Smith Kipp, Rita, *The Early Years of a Dutch Colonial Mission: The Karo Field*. Ann Arbor: University of Michigan Press, 1990.

Snowden, Robert, and Barbara Christian, eds., *Patterns and Perceptions of Menstruation: A World Health Organization International Collaborative Study.* London: Croom Helm, 1983.

Song, Eun Young, "1970 nyŏndae yŏgamunhwa-wa taejungsobi-ŭi chŏngch'i" (1970년대 여가문화와 대중소비의 정치) [authorized translation: "The Politics of Leisure Culture and Mass Consumption in 1970s"], *Hyŏ ndaemunhak-ŭi yŏ n'gu* (현대문학의 연구) ["Journal of Korean Modern Literature"] 50, 2013: 39–72.

Soto, Hernando de, *The Other Path: The Economic Answer to Terrorism.* New York: Basic Books, 1989.

Sousa, M. Luísa, "Colonial Centres and Peripheries: Low-cost Roads and Portuguese Engineers in the 1950s," in: Fari, Simone, and Massimo Moraglio, eds, *Peripheral Flows: A Historical Perspective on Mobilities between Cores and Fringes.* Newcastle upon Tyne: Cambridge Scholars Publishing, 2016: 169–188.

Stanley, Henry Morton, *The Congo and the Founding of Its Free State: A Story of Work and Exploration*, Vol. 2. Cambridge: Cambridge University Press, 2011 (orig. 1885).

Stornig, Katharina, *Sisters Crossing Boundaries: German Missionary Nuns in Colonial Togo and New Guinea, 1897–1960.* Göttingen: Vandenhoeck and Ruprecht, 2013.

Straeten, Jonas van der, "Borderlands of Industrial Modernity: Explorations into the History of Technology in Central Asia, 1850–2000," *Technology and Culture* 60 (3), 2019: 659–687.

Straeten, Jonas van der, *Capital Grids: A Global History of Electricity in East Africa.* New York: Palgrave Macmillan, forthcoming.

Straeten, Jonas van der, and Heike Weber, "Technology and Its Temporalities: A Global Perspective," in: Carnino, Guillaume, et al., eds, *Global History of Techniques (19th–21st Centuries).* Turnhout, Belgium: Brepols, forthcoming 2022.

Straeten, Jonas van der, and Mariya Petrova, "Mud Bricks in a Concrete State: Building, Maintaining and Improving One Own's House in Soviet Samarkand, 1957–1991," in: Krebs, Stefan and Heike Weber, eds, *The Persistence of Technology: Histories of Repair, Reuse and Disposal.* Bielefeld: transcript, 2021: 111–137.

Stronski, Paul, *Tashkent: Forging a Soviet City, 1930–1966.* Pittsburgh, PA: University of Pittsburgh Press, 2010.

The Sunday Post (Nairobi), March 9, 1969b.

Tanganyika Herald, May 30, 1933.

Tanganyika Territory, *Report of the Central Development Committee.* Dar es Salaam: The Government Printer, 1940.

Thompson, Edward P., *The Making of the English Working Class.* London: Gollancz, 1963.

Tinn, Honghong, "From DIY Computers to Illegal Copies: The Controversy over Tinkering with Microcomputers in Taiwan, 1980–1984," *IEEE Annals of the History of Computing* 33 (2), 2011: 75–88.

Tjoa-Bonatz, Mai Lin, and Mikael Hård, "Creole Objects and Techniques: Gold Mining, Gold Panning and Gold Working in Colonial Indonesia," *Baessler-Archiv* 67, 2021: 67–94.

Töpperwien, Annemarie, *"Seine Gehülfin": Wirken und Bewährung deutscher Missionarsfrauen in Indonesien 1865–1930.* Cologne: Köppe, 2002.

Tremlett, Horace, *With the Tin Gods.* London: John Lane the Bodley Head, 1915.

Turner, John, F.C., *Housing by People: Toward Autonomy in Building Environments.* London: Marion Boyars, 1976.

Urry, John, *The Tourist Gaze: Leisure and Travel in Contemporary Societies.* London: Sage, 1990.

Vansina, Jan, *Oral Tradition as History.* Madison, WI: University of Wisconsin Press, 1985.

Vasilyeva, Zinaida, *From Skills to Selves: Recycling "Soviet DIY" in post-Soviet Russia.* Neuchâtel: University of Neuchâtel, 2019.

Vorlaufer, Karl, *Koloniale und nachkoloniale Stadtplanung in Dar es Salaam: Gesellschaftspolitische Zielvorstellungen und städtebauliche Ideen in ihrem Einfluß auf die Raumstruktur einer tropischen Großstadt.* Frankfurt am Main: Frankfurt University, 1970.

Vosotras (Buenos Aires), Vol. 6, 1940.

Vostral, Sharra L., *Under Wraps: A History of Menstrual Hygiene Technology.* Lanham, MD: Rowman & Littlefield, 2008.

Wang, Hsiu-Yun, "Postcolonial Knowledge from Empires: The Beginnings of Menstrual Education in Taiwan, 1950s–1980s," *East Asian Science, Technology and Society* 11 (4), 2017: 519–540.

Waterson, Roxana, *The Living House: An Anthropology of Architecture in South-East Asia.* Singapore: Oxford University Press, 1991.

Weber, Max, "Science as a Vocation," in: Gerth, H.H., and C. Wright Mills, eds, *From Max Weber: Essays in Sociology.* New York: Oxford University Press, 1958: 129–156.

Wenzlhuemer, Roland, "The Telegraph and the Control of Material Movements: A Micro-study about the Detachment of Communication from Transport," *Technology and Culture* 58 (3), 2017; 625–649.

Whitcombe, Elizabeth, *Agrarian Conditions in Northern India, Vol. 1: The United Provinces under British Rule, 1860–1900.* Berkeley, CA: University of California Press, 1972.

Winner, Langdon, "Do Artifacts Have Politics?" *Daedalus* 109, 1980: 121–136.

Wright, Georg Henrik von, *Explanation and Understanding*. London: Routledge & Kegan Paul, 1971.

Yi, Jong-Hyun, *History of Korean Modern Retailing: Repressed Consumption and Retail Industry, Perceived Equality and Economic Growth*. Leiden: Brill, 2016.

Yŏhaksaeng (여학생) ["Female Students"] (Seoul), 1975–1985.

Yŏsŏngdonga (여성동아) ["East Asian Women"] (Seoul), 1968–1981.

INDEX[1]

[1] Note: Page numbers followed by 'n' refer to notes.

M. Hård, *Microhistories of Technology*, Palgrave Studies in the History of Science and Technology,
https://doi.org/10.1007/978-3-031-22813-1

FSC
www.fsc.org
MIX
Papier | Fördert
gute Waldnutzung
FSC® C083411

Kolibri 360 GmbH
Ferdinand-Jühlke-Straße 7
99095 Erfurt, Deutschland
produktsicherheit@kolibri360.de